ISBN 978-1-333-70107-9
PIBN 10537479

This book is a reproduction of an important historical work. Forgotten Books uses state-of-the-art technology to digitally reconstruct the work, preserving the original format whilst repairing imperfections present in the aged copy. In rare cases, an imperfection in the original, such as a blemish or missing page, may be replicated in our edition. We do, however, repair the vast majority of imperfections successfully; any imperfections that remain are intentionally left to preserve the state of such historical works.

1 MONTH OF
FREE
READING

at
www.ForgottenBooks.com

By purchasing this book you are eligible for one month membership to ForgottenBooks.com, giving you unlimited access to our entire collection of over 700,000 titles via our web site and mobile apps.

To claim your free month visit:

www.forgottenbooks.com/free537479

RUSK'S
Illustrated Guide

TO THE

 atskill Mountains

WITH

MAPS AND PLANS.

SAMUEL E. RUSK, PUBLISHER,
CATSKILL, N. Y.

Price, with Prof. A. Guyot's Map of the Catskills, 75 Cts.
Price, complete. without the above Map, 25 Cts.

AN

ILLUSTRATED GUIDE

TO THE

CATSKILL MOUNTAINS;

WITH

MAPS AND PLANS.

———

BY

SAMUEL E. RUSK.

———

A BOOK OF FACTS.

———

SAMUEL E. RUSK, PUBLISHER,
CATSKILL, N. Y.

F. H. WEBB,
Printer and Stereotyper,
HUDSON, N. Y.

F₁₂₇
C₃F₃

TO THE READER.

The number of visitors to the Catskill Mountains has increased ten fold during the past fifteen years. A wide-spread desire has thus been created for accurate Information concerning the different localities; how to reach them; where to go, and what to see.

The author having, with Professor Guyot, and alone, made surveys and measurements of the Mountains, and having been acquainted with the place and its inhabitants for many years, has had an opportunity for obtaining much information valuable for a reliable Guide.

The work throughout is intended to be of a practical character. The book is made for every-day use, in reaching the Catskills, and in walking and driving among them.

The few poetical extracts inserted give perfect descriptions of the scenery to which they are applied ; in fact, some of them were written concerning these particular places.

It is well known that guide books are too frequently prepared in the interest of those who pay the most money for advertising, thus often deceiving the purchaser. Such matter is presented in these pages as the author supposes the public desire, entirely independent of any preference or influence of any house or locality. Every advertisement contained herein is in the form of an advertisement.

The map of the Mountains, found in one edition of the Guide, has just been prepared from accurate surveys made especially for the purpose by Professor A. Guyot, of the College of New Jersey. It is the only map of the Catskill Mountains that has ever been made.

The other map, embracing the section where most hotels and boarding-houses are centered, has been made from actual surveys, to show the location of all houses and objects of interest. with the roads and paths leading to them.

Most of the illustrations have been drawn from Nature and engraved especially for this Guide, and they are faithful representations of the subjects. No previous engravings have been made of many of these scenes.

Words printed in **full face type** call attention to prominent features in various paragraphs. The index will facilitate reference to anything mentioned.

The author acknowledges his indebtedness to Professor Guyot for the use of all of the figures showing the altitudes of various places in the Mountains, which are here given to the public before their publication by him ; also for many other favors, tending to make this little work more accurate and complete.

This Guide will be frequently revised, so that each edition may be relied on as correct to date.

Hotel and Boarding-house Directory.

Parties corresponding with any of these houses in reference to board will confer a favor by mentioning RUSK'S GUIDE. See index, in back of book, for advertisements.

CAIRO.

		Post-office	Accommodation
Walters' Hotel,	Walters Brothers,	Cairo	50

Telegraphic address, Cairo, N Y

CATSKILL.

Grant House,	Grant & Cornell,	Catskill,	300
Gunn's Hotel,	Enos Gunn,	Catskill.	75
Irving House,	H. A. Person,	Catskill.	100
Olney House,	George R. Olney,	Catskill, Box 338.	30
Prospect Park Hotel,	Prospect Park Hotel Co.,	Catskill.	400
Salisbury House,	James Salisbury,	Catskill, Box 366.	25
Summit Hill House	P. M. Goetchius,	Catskill.	150

Telegraphic address, Catskill, N. Y.

CATSKILL MT. HOUSE.

Catskill Mountain House,		Catskill, N. Y.	400

Telegraphic address, Catskill Mt. House, N. Y.

HAINES' FALLS REGION.

Clifton House,	E. F. Haines,	Catskill, Box 71	80
Glen Cottage,	Owen Glennon,	Catskill	45
Haines' Corner,	Miles A. Haines.	Tannersville.	40
Haines' Falls House,	C. W. Haines,	Catskill.	80
High View House,	Mrs. G. W. Reed.	Catskill.	25
Maplewood,	E. H. Layman,	Tannersville.	25
Roe's Cottage,	Hiram Roe,	Catskill.	20
Shady Grove,	John O'Hara,	Catskill.	35
The Vista,		Catskill.	25

Telegraphic address, Haines' Falls, N. Y.

HENSONVILLE.

Griffin's Rural Retreat,	O. S. Griffin,	Hensonville.	25

Telegraphic address, Hensonville, N. Y.

HUNTER.

Central House,	J. Rusk & Son,	Hunter.	65
Hunter House,	M. C. Van Pelt,	Hunter.	80

Telegraphic address, Hunter, N. Y.

KISKATOM.

		Post-office.	Accommodation
Half-Way House,	David Bloom,	Catskill.	50

Telegraphic address, Catskill, N. Y.

LAUREL HOUSE REGION.

| Laurel House, | J. L. Schutt, | Catskill. | 125 |

Telegraphic address, Laurel House, N. Y.

OVERLOOK MT. HOUSE.

| Overlook Mt. House, | James Smith, | Woodstock. | 200 |

Telegraphic address, Overlook Mt. House, N. Y.

PALENVILLE.

Hawver House,	P. Hawver,	Palenville.	40
Maple Grove House,	Philo Peck,	Palenville.	100
Palenville Hotel,	Peter Burger,	Palenville.	40
Pine Grove House,	C. DuBois,	Palenville.	75
Pleasant View House,	George Haines,	Palenville.	18
Sunny Slope House,	P. H. Scribner,	Palenville.	40
The Winchelsea,	Theo. C. Teale,	Palenville.	40
	O. Adsit,	Palenville.	16
	T. N. Lawrence,	Palenville.	25

Telegraphic address, Palenville, N. Y.

TANNERSVILLE.

Blythewood,	Mrs. Alex. Hemsley,	Tannersville.	60
Cascade House,	G. N. Eggleston,	Tannersville.	45
Elm Cottage,	Miss L. A. Craig,	Tannersville.	15
Fairmount House,	William Wooden,	Tannersville.	40
Meadow Brook House,	A. Stimpson Haines,	Tannersville.	35
Mountain Home,	Aaron Roggen,	Tannersville.	85
Mountain Summit House,	S. S. Mulford,	Tannersville.	75
Tannersville Cottage,	George Campbell,	Tannersville.	20
	Frank Eggleston,	Tannersville.	15

Telegraphic address, Tannersville, N. Y.

WINDHAM.

Osborn House,	O. R. Coe,	Windham, L. Box B.	40
Windham House,	Sherman Munger,	Windham.	65
	John Soper,	Windham.	30
	L. S. Graham,	Windham.	15

Telegraphic address, Windham, N. Y.

*ALTITUDES

Above mean tide in New York harbor of some of the principal peaks and points of interest in the Catskill Mountains, as measured—

By A. Guyot.

Hunter Mountain	4,040
Black Dome	4,003
Black Head	3,945
Big Westkill Mountain	3,896
Stony Mountain. east end	3,841
Mink Mountain	3,807
High Peak	3,664
Schoharie Peaks, west and highest peak	3,650
Rusk Mountain	3,624
Indian Head, west and highest peak	3,581
Windham High Peak	3,534
Round Top	3,500
North Mountain, West Peak	3,440
North Mountain, East Peak	3,285
North Mountain, The Outlook	3,108
Ashland Pinnacle	3,420
Plaaterkill Mountain, approximate	3,280
Eastkill Mountain, approximate	3,190
Colonel's Chair, highest, Barometer Station B	3,165
Colonel's Chair, north end, Barometer Station A	3,037
Overlook Mountain	3,150
East Jewett Mountain	3,146
Overlook Mountain Honse	2,978
Pisgah Mountain	2,905

Indian Pass, highest point of trail between Plaaterkill Mountain and
Indian Head............ 2,694

Mink Hollow, highest point of road...................... 2,629

Parker Hill............... 2,545

South Mountain............ 2,497

Clum Hill........... 2,372

Catskill Mountain House 2,225

Catskill Lakes 2,138

Point of Rocks 2,128

Sunset Rock, on South Mountain. 2,115

Laurel House   . ...,....... 2,038

Grand View House........... .:................. 1,970

The Vista..................................... ....-............ 1,932

Tannersville, Mountain Home 1,926

Plaaterkill Falls, Dibbles   . . ..-..... 1,855

Stony Clove Notch, approximate   ....,....... ... 1,700

Hunter, J Rusk & Son's.. . .   ..,..... 1,609

Windham 1,510

Lexington ...,.............. ....,..... 1,320

Sleepy Hollow.................................. ... . 1 290

Prattsville ....,.................   1,164

Kiskatom, toll-gate 687

Palenville, Palenville Hotel ....-... 680

Woodstock 594

---

*By special arrangement with Professor Guyot this list of altitudes is first
given to the public through this GUIDE, which is copyrighted Parties are
cautioned against publishing any part thereof without permission.

Hunter Mountain is the highest point in the Catskills.   In the South Catskills is one mountain higher—Slide Mountain, reaching an altitude of 4205 feet.

SANTA CRUZ FALLS.

THE CATSKILLS.

For more than half a century the Catskill Mountains have been visited by many who have sought a few weeks summer recreation. While, until about 1865 they might have been counted by hundreds, since that time each season has added to the number, so that now thousands annually visit this region.   The hotel and boarding-house accommodations have kept pace with the increasing influx of people.

Many things combine to attract the health and pleasure seekers to this section: It is so near New York and easy of access by cars and steamers of the Hudson.   The atmosphere is pure and invigorating and the temperature on the mountains is fifteen degrees lower than at New York.   There are always cool breezes.   The place is free from chills and fever, malaria, and hay fever ; it affords relief from these things.   Owing to the altitude of the mountains, a change of climate is experienced equal to that of going to a much more northerly latitude without the increased elevation.   In a small compass is a varied combination of magnificent scenery, and panoramic views include a large section of the Hudson valley and extend across it into six states.

There is a growing custom of remaining among the mountains later in the autumn than was formerly done ; and this is a commendable practice, for October brings mountain beauties and pleasures that mid-summer does not possess.

PRICE OF BOARD.

The price of board in the Catskills varies from five dollars a week to three dollars and a half a day.   The large houses provide comforts and conveniences equal or superior to similar priced houses in many other parts of the country, while a large number of smaller ones give perfectly satisfactory accommodations at from eight to ten dollars a week.   The price at some houses is frequently higher than at some other similar ones, on-account of being in a more desirable locality.

ROUTES TO THE CATSKILLS.

THE PRINCIPAL WAY.

The most direct route to the greater part of the Catskill Mountains is by the way of Catskill. The village is situated on the west bank of the Hudson River, at the junction of the Catskill Creek. It is one hundred and ten miles from New York, and thirty-three miles from Albany.

TO CATSKILL FROM NEW YORK.

From New York, Catskill may be reached by the New York and Albany Day Line of Steamers, C. Vibbard and Daniel Drew; by the Hudson River Railroad; and by the Catskill Night Boats, Escort and C. Vanderbilt.

The **Day Boats** leave New York every morning, except Sundays, throughout the summer, landing at **Catskill Point** in the middle of the afternoon. THE POINT is the name of the dock built out a short distance into the Hudson where it is joined by the Catskill Creek. Those who desire to enjoy a delightful sail, and to have an opportunity for viewing the varied scenery along the "River of the Mountains," as the Hudson was called by the Spaniards, will make the journey by these palatial steamers. A pocket map, entitled "The Hudson by Daylight Map," with descriptive pages, will be useful to the stranger, by showing prominent residences, historic landmarks, **and** other interesting objects on the banks of the river. **The** fare from New York to Catskill is one dollar and fifty cents. (See index in back of book for time-table.)

The numerous trains on the **Hudson River Railroad** stop at **Catskill Station,** making the time from New York from three and a half to four hours. Catskill Station is on the opposite side of the river from The Point, and is connected therewith by ferry. The ferriage is thirteen cents. Parties can usually leave New York by a train as late in the day as between eleven and twelve o'clock, and yet reach Catskill as early as the boats do. The summer fare is two dollars and eighteen cents ; being less than in winter. (See index for time table.)

The **Catskill Night Boats** leave New York every evening, except Sundays, and reach Catskill early the following morning. They have very comfortable accommodations ; thus assuring a good night's rest on the cool water. Sometimes they land at The Point, and at other times up the Creek, near the business center of the village. Full particulars may be learned by consulting a time-table in this book. (See index.) The fare, including berth, is one dollar.

TO CATSKILL FROM ALBANY.

Three lines of conveyances may be used to reach Catskill from Albany. The Day Boats previously mentioned, which leave Albany in the morning and arrive at Catskill before noon ; the trains of the Hudson River Railroad, which make the trip in about an hour ; and the steamer City of Hudson, which leaves Albany in mid-afternoon and reaches Catskill in the evening. The fare by the boats is fifty cents, and by the cars seventy-six cents. (See index for time-tables.)

CONVEYANCE FROM CATSKILL TO THE MOUNT-AINS.

Conveyances, in great variety and abundance, are always found in waiting at Catskill on the arrival of the boats and cars. Many of them run from particular hotels and boarding-houses ; others convey parties to any of the houses in certain localities ; while the livery-men of the village will provide vehicles for any place that may be desired. The omnibuses are present to carry to the village hotels.

People who engage board before going to the Mountains— the greater number do so—usually arrange at the same time for their conveyance to the house. This has been found to be a desirable plan : for, both time and the inconvenience sometimes incident in securing a satisfactory carriage are thereby avoided.

In the description of each particular locality mentioned in these pages, under its appropriate head, may be found the special details concerning conveyance to it. (See index for names of places.)

For the Catskill Mountain House, the Laurel House, Palen-ville, Haines' Falls, Tannersville and Hunter the road crosses the Catskill Creek, in the village.

Three miles and a half out, over a hilly region of varied beauty, the Cauterskill Creek is crossed in a deep valley. In ascending the hill on the west side, there may be seen, down toward the creek, half a mile to the left, an old, low, stone house. From this house, in 1781, David Abeel and his son Anthony were taken prisoners by a band of Indians and tories, and carried to Canada ; stopping for one or two nights in the Old Indian Fort, between High Peak and Round Top.

Half a mile farther on is **Glenwood Hotel,** and another half mile reaches the **Mountain Retreat House**—both located in Kiskatom—and at an elevation above the Hudson of some three hundred feet. The name Kiskatom is said to be of Indian origin, meaning hickory tree or nut. These trees abound in this region.

Just beyond this point roads diverge : the one to the right leads to the Mountain House and Laurel House, while the left hand one passes up through the Cauterskill Clove.

The road out of the upper end of Main Street leads, by way of Jefferson, to Leeds, Cairo and Windham. At Leeds, three miles from the village, the road passes over a picturesque old stone bridge.

THE RONDOUT ROUTE.

The Overlook Mountain House, Hunter, and some places in the south-western part of the Mountains, are most conveniently accessible from the Hudson by the way of Rondout. Rondout is a part of the city of Kingston, situated on the west bank of the Hudson, opposite Rhinebeck. Its distance from New York is eighty-nine miles, and from Albany fifty-three miles.

TO RONDOUT FROM NEW YORK.

Trains on the Hudson River Railroad; and the New York and Albany Day Boats, stop at Rhinebeck. The boats leave New York in the morning and land at about two o'clock. A ferry connects with Rondout. The fare by the cars is one dollar and seventy-six cents, and by the boats one dollar and twenty-five cents. The ferriage is thirteen cents.

By way of the Erie and Wallkill Valley Railroads, Rondout may be reached direct. The fare is one dollar and eighty-eight cents. There is, also, a line of Night Boats which leave New York late every afternoon, except Sundays, and arrive at Rondout early the following morning. The fare by these boats is seventy-five cents.

TO RONDOUT FROM ALBANY AND FROM THE EAST.

The facilities for reaching Rondout from Albany are by the Hudson River Railroad ; and the Day Boats, which land between twelve and one o'clock. The car fare is one dollar and eighteen cents, and the fare by boats is seventy-five cents, to Rhinebeck.

The western terminus of the Rhinebeck and Connecticut Railroad being at Rhinebeck, parties from the East can conveniently make that their line of travel.

TO THE MOUNTAINS FROM RONDOUT.

Having arrived at Rondout by any of the several converging lines mentioned, passage to the mountains is continued by the Ulster and Delaware Railroad.

The places that are mentioned in these pages, for which this route should be chosen to reach, may be found, with descriptions of the same, under suitable headings. Here, also, may be learned the proper railroad stations to leave and other particulars.

PROSPECT PARK HOTEL, CATSKILL.

TO CAIRO

HOPE HOLLOW

CATSKILL

CREEK

S

Cauterskill
Cr.

TO CAUTERSKILL

TO THE   MOUNTAINS

TO SAUGERTIES

PLAN OF

CATSKI

BY SAMUEL E. R

1879.

TO SAUGERTIES

PLAN OF
CATSKILL
BY SAMUEL E. RUSK.
1879.

:r skill
r.

TO C.

BRIDGE

RAM's HORN

THE POI

KILL

I MILE

CATSKILL VILLAGE AND VICINITY.

The name Catskill was derived from the Dutch **Katzkill,** meaning cats' river or stream ; probably so called from the many panthers or wild cats that formerly infested this locality and the mountains bearing the same name.

Only a small part of the village is visible from the Hudson or even from the landing at The Point. A narrow ridge of land lies between the river and the Catskill Creek ; and it is on the western slope of this ridge, along the creek, where the greater part of the village is situated. From the river, however, a few comfortable looking residences appear along the top of the ridge. The one large building overlooking the Hudson from the southern end of the ridge, and which is so conspicuous from the landing and from quite a long distance along the river, is the **Prospect Park Hotel.**

The sketch on another page will convey an idea of the **Mountains** as they appear from Catskill. A distance of ten miles reaches the nearest point of their base, and an elevation of some seven hundred feet above the Hudson. From this base they rise, almost perpendicularly, thousands of feet in their perpetual sublimity.

While, midway between the northern and southern limits of the lofty peaks forming the front of the range, the **Catskill Mountain House** stands in full view, the **Overlook Mountain House** is barred from sight from this point by the dark clothed summit of Overlook Mountain on whose southern slope it is located; but, from a short distance down the river, it stands out in as clear relief against the sky as the Catskill Mountain House does from here.

There, on his back, lies the **Old Man of the Mountains,** seemingly unmindful of the many storms that sweep his rocky bed. The tri-topped Indian Head forms his rugged visage. pillowed between Schoharie Peak and the long slope of High Peak; the Plaaterkill Mountain, his high heaved chest; while the Overlook Mountain outlines his drawn up knees.

The Cauterskill Clove and Plaaterkill Clove are the only deep gorges cut through the eastern side of the mountains. No finer view of the Hudson valley can be had than that from the summit of High Peak. The maps will show the arrangement of the Mountains.

Catskill being the grand gate-way to the mountain region. and possessing within its precincts so many attractions delightful to the summer pleasure seeker,—it is not surprising that many choose to view the varied mountain scenery from this stand-point. Its locality, directly on the great Hudson River thoroughfare, provides convenient access to New York for those who can leave the city but a day or two at a time.

There are special **conveyances** from the **Prospect Park Hotel** and the **Grant House** to the boats and cars. From the **Irving House** and **Gunn's Hotel,** located on Main Street in the business center of the village, omnibuses meet all boats and trains.

Besides the hotels, Catskill village contains some private boarding-houses. There are six churches, two banks, an opera-house, an academy, two weekly newspapers, a paper mill, a foundry and a woolen factory. There are numerous stores with various lines of goods. Catskill being the shire-town of the county, the court-house is in this village.

The **Post-office** is centrally located at 244 Main Street. There are three or four mails each day between this place and New York. This office receives and forwards nearly all the mail of Greene county. It is also a **Money Order Office.**

The Western Union **Telegraph Office** is at 275 Main street.' **Money Transfers** are made by telegraph between this and all other large Western Union offices. There is also a summer office at the Grant House, on the Catskill, Cairo, and Windham line. An office of the American **Express** Company is at 266 Main street.

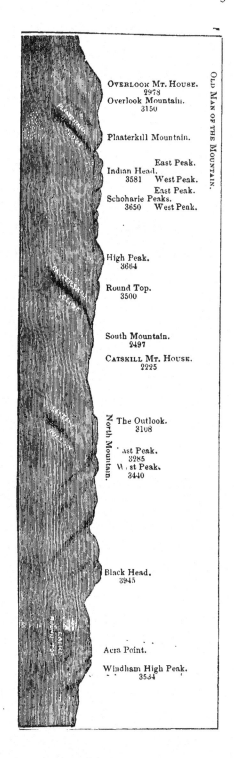

OLD MAN OF THE MOUNTAIN.

OVERLOOK MT. HOUSE.
2978
Overlook Mountain.
3150

Plaaterkill Mountain.

East Peak.
Indian Head.
3581   West Peak.
East Peak.
Schoharie Peaks.
3650   West Peak.

High Peak.
3664

Round Top.
3500

South Mountain.
2497

CATSKILL MT. HOUSE.
2225

North Mountain.

The Outlook.
3108

ast Peak.
3285
W . st Peak.
3440

Black Head.
3945

Acra Point.

Windham High Peak.
3534

Sixty thousand brook and salmon **trout** were procured and put in the creeks and ponds of the towns of Catskill and Cairo in 1876–7 ; and Greene county now has, in successful operation, an establishment for hatching trout and other fish and restocking all the lakes and streams of the county. Many thousands of young trout have thus been already supplied. The establishment is located on a fine trout stream at Palenville. (See index for "Greene County Fish Hatchery.")

For **boating**, no better places need be desired than the Hudson River at this point, and the Catskill Creek. There are plenty of row boats and small sail boats obtainable.

JEFFERSON HEIGHTS

Is a part of Catskill. It is a mile from the center of Catskill village. On this plateau are located the **Grant House** and some other smaller boarding houses.

WALKS ABOUT CATSKILL AND VICINITY.

The fine river view and view of the mountains from the grounds around the Prospect Park Hotel claims this as among the first walks to be made in Catskill. The residence of Church, the artist, appears on a high hill which rises from the eastern bank of the Hudson almost opposite.

Admirably located on the hill in the upper part of the village are the residence and studio of the late Thomas Cole, N. A., of "The Course of Empire" fame. His "A Lake with Dead Trees" and "The Falls of the Cauterskill" were painted after a visit to the Catskills.   That he was greatly delighted to be among the mountains is evinced by the following fragments from his pen:

> Friends of my heart, lovers of Nature's works,
> Let me transport you to those wild blue mountains
> That rear their summits near the Hudson's wave.

and:

> Oh, for an hour
> Upon that sacred hill, that I might sleep,
> And with poetic fervor wake inspired!
> Then would I tell how pleasures spring like flowers
> Within the bosom of the wilderness.

Two hours or less will suffice for a walk up to, and around the Grant House and return.

Three-fourths of a mile above the Grant House is **Hope Hollow** or **Austin's Glen,** in whose cool recesses a half-hour may be pleasantly spent.   This Hollow was the course of a railroad in their earlier days.   A spring of cold water furnishes drink for the thirsty.

DISTANCE FROM CATSKILL BY THE MOST USUAL ROUTE

TO

Acra	13
Ashland	30
Cairo	10
Catskill Mountain House	12
Cauterskill Clove	10
Durham	22
East Durham	15
East Windham	19

PALENVILLE.

The hamlet of **Palenville** is located directly at the entrance of the Cauterskill Clove and is the most western part of the township of Catskill. It was named from a family by the name of Palen, who built three tanneries here; first, a small one, and soon after, two large ones—all dating near the beginning of the present century. The enterprise was successfully carried on for many years. There is yet a large tannery in operation on the site of the one last built by the Palens,—the only one remaining anywhere in this region. Where the other large one was located, there is now a woolen factory which was built several years since. Half a century ago, Greene county produced more leather than all the rest of the State of New York

The base and sides of the mountains above Palenville are dotted with many quarries, producing large quantities of paving and building stones, which are shipped to various cities throughout the United States.

Palenville is but nine miles from Catskill and its general elevation is seven hundred feet. The first summer boarders here were artists, who found this an admirable region in which to obtain choice studies,—from amid such varied mountain wildness as is seldom found more alluring than in the Cauterskill Clove and its surroundings. There are now many first-class summer boarding-houses here, accommodating from a dozen to a hundred people each, and they are well patronized. **The** place is certainly a pleasant one.

The time of day at Palenville is the same as at New York City, for the line of longitude three degrees east from Washing-ton or seventy-four degrees west from Greenwich, passes through both places.

Stages and private conveyances meet parties upon arrival at Catskill, and the drive out, over good roads, is enjoyable. The stage fare is one dollar, which includes baggage. There are plenty of conveyances to be had for **drives** to the places of interest in the vicinity; such as the Catskill Mountain House, through the Cauterskill Clove, Haines' Falls, Kaaters-kill Falls, and Sleepy Hollow. The points here mentioned are so located that a round trip of fifteen miles will reach all of them, but such a trip would not allow sufficient time to properly see them in detail.

There are two daily **Mails ;** the post-office being at the store of Charles H. Teale.

A Western Union **Telegraph Office,** with direct wires to New York, is located in the post-office.

In the centre of the hamlet is a Union **Church.** Measures are being taken to build an Episcopal Chapel in the place.

The Greene County Fish Hatchery is located in Palenville. (See description elsewhere.) This is a novel attraction, espe-cially for visitors who are interested, but who have not hitherto had an opportunity to examine the science and art of piscicul-ture in its practical working.

Near the upper part of the Cauterskill Clove is a small stream, emptying into Santa Cruz Creek, on which are **Bridal Veil Falls**—well seen from near The Winchelsea after prolonged rains.

WALKS ABOUT PALENVILLE AND VICINITY

DISTANCES FROM THE POST-OFFICE.

MILES.

Artists' Grotto and La Belle Falls, in Cauterskill Clove... 1

Catskill Mountain House, by path via Moses' Rock...... 3

Drummond Falls 1

Fawn's Leap and Profile Rock, in Cauterskill Clove..... 2

Fish Hatchery.. 1

Haines' Falls, by road.................................. 4

"   "   by path via Haines' Ravine,—including
    Shelving Rock and The Five Cascades............ 4

Kaaterskill Falls and Laurel House, via Cauterskill Clove
    and Kaaterskill Ravine,—including Bastion Falls.... 4

Moses' Rock... 2

Mossy Rock .... $2\frac{1}{2}$

Palenville Overlook..................................... 2

Santa Cruz Falls... 3

Sunset Rock, via Kaaterskill Ravine and Laurel House.. 5

White Fawn Falls and Black Crook Falls................ $\frac{3}{4}$

Many of the points of interest near Palenville are in the Cauterskill Clove and are described under that heading. (See index.)

DRIVES IN THE VICINITY OF PALENVILLE.

DISTANCES FROM THE POST-OFFICE BY THE MOST USUAL ROUTES.

MILES.

Around South Mountain, via Cauterskill Clove, Haines'
    Falls, Kaaterskill Falls, Mountain House, and Sleepy
    Hollow,—round trip................................ 15

DRUMMOND FALLS.

WALK OR DRIVE IN PALENVILLE.

Drummond Falls are one mile south from the Union Church, and but a few yards off of the road leading to Sau·gerties. The path starts on the left from an abrupt turn in the road just beyond the Drummond Falls House. The distance to the Falls from the post-office is, also, one mile ; by the road on the right, which is shaded a considerable portion of the way.

The Fountain Kill is here formed into a foamy cascade of pleasing detail, as it makes a sudden descent of some forty feet. The walls on either side, and a huge mound of red sandstone, which stands near below, give evidence that the erosive power of the water has gradually formed this attractive nook. A little way on, the stream unites with the Cauterskill.

DRUMMOND FALLS.

BLACK CROOK FALLS. WHITE FAWN FALLS.

WALK IN PALENVILLE.

Black Crook Falls and **White Fawn Falls** are both on the Fountain Kill, but a few rods apart. The distance to them, either from Mrs. Hinman's, near the Winchelsea, or from the Union Church, is half a mile.

The path east of Mrs. Hinman's house is first across the fields, and then in a growth of small oaks and pines, amid which the Falls are found. From the road, north of the Church a fourth of a mile, starts the path which leads up the mountain, via Moses' Rock, to the Mountain House. The White Fawn Falls are a little to the left of this path, an eighth of a mile along.

Neither of these Falls are more than a few feet in height. and yet the brook in its changing moods has a fascination which invites one to spend the morning here.

By whichever of the two ways this spot is reached, **the return** is usually made by the other path.

GREENE COUNTY FISH HATCHERY.

WALK OR DRIVE IN PALENVILLE.

The **Greene County Fish Hatchery** is about one mile south from the post-office, and located by the road, so that it may be reached by driving. It was established for the purpose of greatly increasing the supply of trout, and other fine, edible fish, in the numerous streams and lakes of the County. These waters are adapted to the habits of a variety of fish.

The following figures show the work of the Hatchery in stocking the waters for two years :

	1878.	1879.
Brook Trout......................	288,000	200.000
Salmon Trout.....................	38,000	90,000
California Salmon.................		15.000
White Fish.......................		25,000
Black Bass.......................		

It is the intention to continue this undertaking for three years more, supplying about the same or an increased number of young fish as above shown during each year. The work is superintended by A. W. Marks, of the New York State Fishery.

Salmon trout will seldom bite a bait, and are usually caught by trolling, or in nets. California salmon are taken with a fly. and black bass are the best fish for fly fishing that we have. All of the kinds mentioned, except the brook trout, are to be found in the lakes and large creeks instead of the small streams.

An angler might be tempted to use his hook at the Hatchery, where 2,500 full-grown brook trout may be seen together. The crystal water of the winding **Spring Creek,** which flows by, and through the Hatchery, has the peculiar characteristics which the trout naturally seek.

> I come from haunts of coot and hern,
>   I make a sudden sally,
> And sparkle out among the fern,
>   To bicker down a valley.
>   *   *   *   *   *
> I chatter over stony ways,
>   In little sharps and trebles,
> I bubble into eddying bays,
>   I babble on the pebbles.
>   *   *   *   *   *
> I wind about, and in and out,
>   With here a blossom sailing,
> And here and there a lusty trout,
>   And here and there a grayling. —*Tennyson.*

WALK FROM PALENVILLE.

Palenville Overlook is the high point of South Mount-ain on the right of the entrance to Cauterskill Clove, and over-looks Palenville. A small house marks the spot. The distance from the post-office is two miles, and the roads and paths which wind up the rugged steep to it are shown on the map.

Its elevation above Palenville is some fifteen hundred feet and the side of the mountain is so nearly perpendicular that the chief view of the houses below is of their roofs. The view of the Hudson valley is similar to the one from the Mountain House, but less extended.

There is a path from here over South Mountain to the Mountain House—a walk of a mile and one fourth; and also another path along the side of the mountain which intersects the road below the Mountain House.

CAUTERSKILL CLOVE.

No visitor to the Catskills should depart without having **seen** 'he many attractions of the justly celebrated **Cauterskill Clove.** It is a fact to be especially noticed that the greater number of the water-falls in the whole range of the Catskill Mountains are centered around this Clove. The roads from Catskill and Saugerties converge at the entrance, which is at the upper end of Palenville. It is three miles, or a little more, from here to the top of the mountain, and ten miles to Hunter —this road constituting the Hunter Turnpike.

By the bridge, at the entrance of the Clove, **is the** quaint studio of the artist Hall. Opposite is the Palenville **Hotel;** at whose watering tub horses are watered when driving through the Clove. On the south side rises the long and steep slope of High Peak ; while to the north an almost perpendicular wall reaches the height of seventeen hundred feet above the road, at which point is the Palenville Overlook.

Just above the toll-gate, an eighth of a mile along, **is a path** leading down the steep bank to the creek, where are located the **Artists' Grotto** and **La Belle Falls.** It will take but a few minutes to reach these interesting places.

A few rods on from the toll-gate a land-slide extends from the road down to the creek, and on the opposite bank is a high, perpendicular wall of red sandstone. The passage between them is the **Red Chasm.** The projecting point of the top of the mountain above it is the **Point of Rocks.** Just west of this Point is a wild gorge, extending down the side of South Mountain, and partially dividing it, which is known as **The Gulf.** An old road or path, starting from the bridge at the entrance of the Clove, leads along up the side of the mountain and has several diverging branches which reach The Gulf at different elevations.

Soon the road passes the cozy summer cottage of E. T Mason. A short distance more reaches the top of a hill at a turn in the road, to the right of which the creek flows through a defile in the rocky barrier called **Deep Chasm.**

The objects of special interest to be next seen are at **More Bridge,**—a trifle more than a mile on the way,—which spans the Cauterskill. **Church's Ledge** rises close by the road and bridge, and part of its irregular side, seen from a point across the Bridge and a few feet to the right of the road, resolves itself into the features of **Profile Rock. More Falls** are just below the bridge, and above, the water has worn queerish channels and circular cavities into the red sandstone bed of the creek.

At the west end of Church's Ledge is a gorge known as **Hillyer's Ravine.** Half way up the steep, visible side of the mountain it expands into a wide basin. Above are precipitous cliffs, down which dash the waters of a small stream, forming Viola Falls. (See description elsewhere of Viola Falls, Wild Cat Falls, Buttermilk Falls, and Santa Cruz Falls, which are here only mentioned.)

From More Bridge the road follows for a ways close on the

VIEW OF MORE BRIDGE.

north bank of the Cauterskill, whose waters a few rods along come pitching down between the cleft rocks into a dark shaded pool below. This chasm is called **Fawn's Leap ;** tradition saying that a young deer, being pursued by a hunter and his dog, leaped across the gulf and escaped, while the dog fell into the water beneath and was drowned. The spot is so close to the road that but a few minutes will be consumed in stopping to view it.

A small stream enters the Cauterskill above Fawn's Leap, up which are located Wild Cat Falls of a hundred feet in height.

The curved side of South Mountain up to the right, with its many ledges rising one above another forms **The Amphitheater,** best seen in the autumn when the foliage is less dense. At its western extremity the projecting part of the mountain forms Sunset Rock.

At the base, the road passes along a flat where was once quite a hamlet ; but now only the **Old Tannery Ruins** and a few piles of stone on either side mark the sites of the numerous buildings that existed when the mountains furnished plenty of hemlock bark for tanning purposes. Behind the Tannery Ruins is Buttermilk Ravine with a pleasing Cascade visible from the road, while the upper part of the Ravine contains the Buttermilk Falls proper. In crossing the flat, Santa Cruz Ravine may be seen on the left, descending the side of the mountain in an easterly direction.

The road takes a short turn at the end of the flat and gradually ascends to **Lake Creek Bridge.** From here, up, is the really steep part of the mountain road, a mile long, and people usually walk this portion to lighten the horses' burden

A path leads up Lake Creek through the Kaaterskill Ravine to the Kaaterskill Falls and Laurel House one mile distant, passing Bastion Falls on the way. To the left of the Bridge and opposite this path is another path leading down to the Cauterskill and thence up through the ravine to Haines' Falls, a mile and an eighth distant, passing by Naiads' Bath, Shelving Rock, and The Five Cascades.

Dripping Rock is a moss-covered ledge by the side of the road half way up the mountain from Lake Creek Bridge over which flows a small stream of excellent potable water. During the remainder of the ascent the increasing exhilerating effect of the pure mountain atmosphere is especially noticeable. A portion of Haines' Falls may be seen from a ways below Dripping Rock.

Near the top of the mountain is a road to the right, called **Feather-bed Lane,** connecting with the Mountain House road. It is steep and rough and not adapted to driving. A few yards more reaches the top of **The Great Land-slide,** which extends from the road down to the bed of the creek—a descent of nearly five hundred feet. The view from here down through the Cauterskill Clove and beyond to the Berkshire Hills of Massachusetts is much admired and is specially delightful near the sunset hour.

Having here reached the upper end of the Cauterskill Clove, it is but a short distance to the Haines' Falls House.

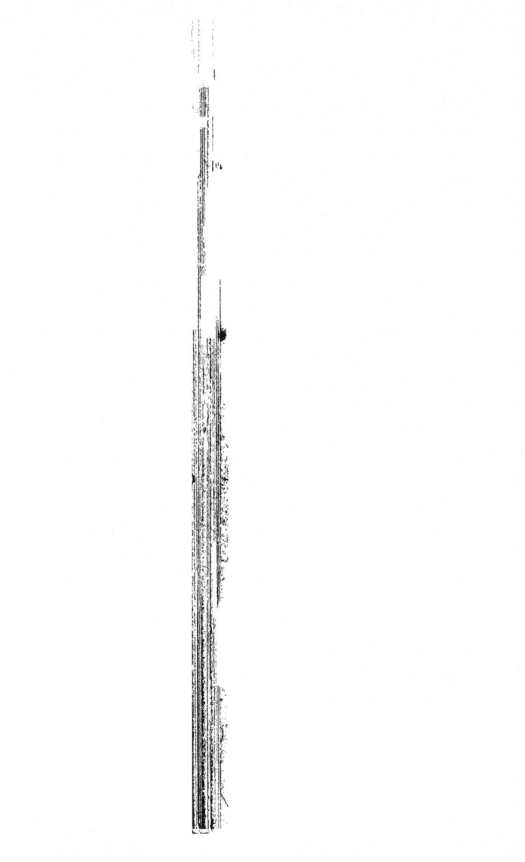

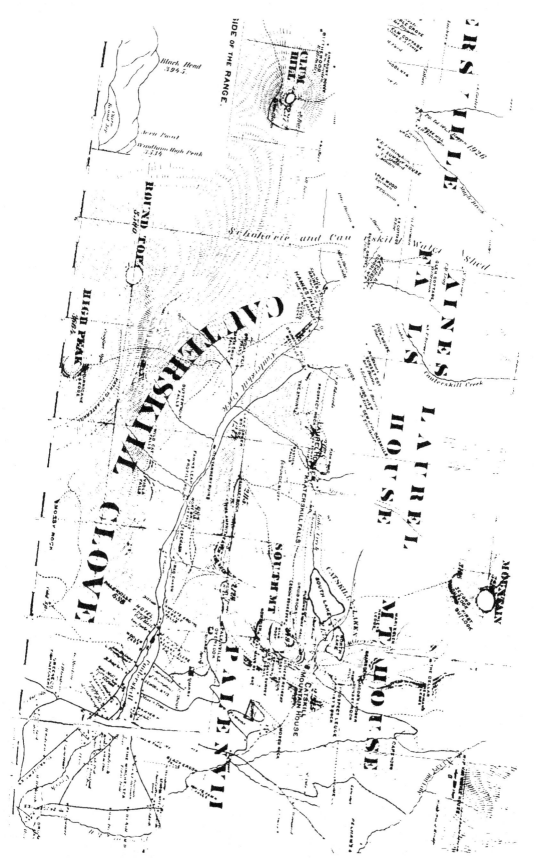

HAINES' FALLS REGION.

The name **Haines' Falls** has been used in three senses. First it was applied particularly to the Falls; afterward, also, to the house of Charles W. Haines, at the Falls; and, more recently, any of the numerous houses within a radius of about a mile are mentioned as being at Haines' Falls. In order to clearly designate which is meant, it has been thought best to use the name Haines' Falls for the Falls only; Haines' Falls House for the house there; and Haines' Falls Region for the locality in general.

The place is thirteen miles from Catskill, situated at the head of the long-time famous Cauterskill Clove, and, consequently, at the front of the mountain range. The view through the Cauterskill Clove extends beyond the Hudson, into Massachusetts and Connecticut. The elevation above the Hudson of points where the houses are placed, varies from one thousand and nine hundred to two thousand and three hundred feet. On the south side the land rises quite abruptly to the summits of High Peak and Round Top, while to the north a more gradual ascent leads to the top of the North Mountain ridge. It is a region of much singular wildness and scenic beauty; and many years ago, before other people had made it their summer home, noted artists sought it out as a rich field of subjects for their canvas. The view of the Hudson valley, as seen from near the Shady Grove House, is excellent.

With the exception of a few hotels that have been in existence for many years, this region was the first to provide summer boarding-houses among the Catskills. There are now many houses, with accommodations for from twenty to a hundred people each. While, during the past fifteen years, the accommodations have been increased forty-fold, there is, almost every year, added provision for the increasing influx of people.

In this region is the Cauterskill and Schoharie water-shed, which sends the water east to the Hudson at Catskill, or, in the opposite direction; flowing by the circuitous route of the Schoharie and Mohawk Rivers to the Hudson above Troy, and following in its course some two hundred and fifty miles to Catskill, where it is then but a dozen miles from its source.

From most houses in the Haines' Falls Region **conveyances** run daily to Catskill to meet the Day Boats each way, and the trains as late as half past three o'clock. For a later hour special arrangements can be made. The regular fare, including baggage, from Catskill is two dollars.

Conveyances may be had for local excursions.

The **Mails** between here and New York are sent and received daily or oftener—Catskill being the post-office.

The Western Union **Telegraph Office,** with direct wires, is at the Haines' Falls House.

There is a Methodist Episcopal **Church,** in which services of other denominations as well as its own are held.

HAINES' FALLS.

in
o
re
a
he
ost
le.
ed,
he
he
nd
to

y-
ay,
ter
are,

and

res,

ices

WALKS ABOUT THE HAINES' FALLS REGION AND VICINITY.

DISTANCES FROM THE HAINES' FALLS HOUSE.

MILES.

Catskill Mountain House, by path via Laurel House	3
" " by road	4
Clum Hill	$2\frac{3}{4}$
Fawn's Leap and Profile Rock, in Cauterskill Clove	2
Haines' Falls	1-16
Haines' Falls Ravine,—through it to Lake Creek Bridge	$1\frac{1}{8}$
High Peak	4
Kaaterskill Falls and Laurel House, by path via Prospect Rock	$1\frac{1}{2}$
Kaaterskill Falls and Laurel House, including Bastion Falls, via the turnpike to Lake Creek Bridge; thence up the Kaaterskill Ravine	2
Kaaterskill Falls and Laurel House, by road	$2\frac{1}{4}$
Old Indian Fort, between High Peak and Round Top	$4\frac{1}{2}$
Overlook Mountain House, by path around the summit of High Peak to Plaaterkill; thence by new road	11
Parker Hill	3
Plaaterkill, by path around the summit of High Peak	6
Prospect Rock	1
Rifted Rocks	$\frac{1}{4}$
Santa Cruz Falls	1
Tannersville	$2\frac{1}{2}$
The Five Cascades, Shelving Rock, Naiad's Bath, Triton Cave, in Haines' Ravine	$\frac{1}{4}$
The Sphinx, or Noah's Ark	$\frac{7}{8}$

To Buttermilk Falls, $1\frac{3}{4}$; thence to Wild Cat Falls, $\frac{3}{8}$;

thence to Viola Falls, $\frac{1}{4}$ .... $2\frac{3}{8}$

There are descriptions of nearly all of the above walks in these pages. (See index.) Where the routes are by direct roads no descriptions are given. The map will show the location of these places and the way to them.

DRIVES IN THE VICINITY OF THE HAINES' FALLS REGION.

DISTANCES FROM THE HAINES' FALLS HOUSE, BY THE MOST USUAL ROUTES.

MILES.

Around South Mountain, via Mountain House, Sleepy
Hollow, Palenville, and Cauterskill Clove,—round trip. 14

Catskill Mountain House............................. 4

Cauterskill Clove,—through it to Palenville............ $3\frac{1}{2}$

Clum Hill.. .... $2\frac{3}{4}$

Hunter 7

Kaaterskill Falls and Laurel House.................. $2\frac{1}{2}$

Overlook Mountain House, by new road via Plaaterkill ...$14\frac{1}{2}$

Parker Hill .... $4\frac{1}{2}$

Plaaterkill .... $9\frac{1}{2}$

Sleepy Hollow, where Rip Van Winkle slept!........... $6\frac{3}{4}$

Stony Clove....................................... 9

Tannersville...................................... $2\frac{1}{2}$

(See index to find description of above places.)

TO HAINES' FALLS AND UNDER THEM.

At the east side of Haines' Falls House a road leads down to **Haines' Falls,** which are but a few rods distant from the house, on the Cauterskill Creek. Passing through a gate, a payment of twenty-five cents each is required. No charge is made for subsequent visits during the same season.

A short, well defined path, to the right, leads through a growth of laurels on the brink of a precipice to the top of the Falls. Here, from the outer extremity of a pendent rock at the side of the Falls, the view disclosed is extraordinarily wild. The main sheet of water pours over the projecting rock into the narrow, deep-cleft gorge, a hundred and fifty feet ; while on the south side the waters of the beautiful **Spray Fall** make a sheer descent of a hundred and sixty feet.

Just before passing over the Falls, the water descends a fifteen foot ledge. Its long continued action has worn a number of circular holes into the solid rock at the foot of the ledge. The principal cavity is several feet deep. Many sketches and paintings have been made of the view from the top of this ledge down through the ravine.

Sometimes in winter a hollow icicle, an hundred feet in length, hangs from the top of the Falls ; and the water which flows with muffled sound down through the interior forms, underneath, a huge cone of glistening ice.

To go **under the Falls,** the path to the left of the gate should be followed. It is but a few steps to the first long flight of stairs. At their foot is a natural gateway, formed by **two** trees, one of which has grown so that the wall of rock is imbedded in its side. A few yards beyond this gate-way is **Crystal**

Spring ; of pure, ice-cold water, which emerges from underneath the massive ledge of rock that has just been descended
Several more flights of stairs are passed,—from which glimpses
of the falling water may be had through the foliage,—before
reaching the bottom of the Falls.

The accompanying illustration was made from the foot of the
Falls, and it is from this point that the most comprehensive
view is obtained of their wildness and sublimity.   While visitors
are down here, the ordinary stream of water is increased from
a pond above, thereby adding to the beautiful spectacle.   At
such times it is avdisable to be down the stream as far as the
stairs, unless one does not mind being drenched by the cloud
of spray.   Farther down the gorge are **The Five Cascades**
(See description.)   The bottom of the Cascades is less than a
fourth of a mile from the top of the Falls ; yet, in this short distance, the stream descends nearly five hundred feet.

The Falls and Cascades were the subjects of numerous
paintings, many years ago, by such artists as Kensett, Cassilear.
Cole and Durand,   In those days ropes and ladders were daily
used in descending and ascending ; but now, good, safe stairs
render access comparatively easy for the artists, with their
implements, as well as for others ; and down here, Gifford and
M'Entee have often been.

THE FIVE CASCADES, IN HAINES' RAVINE.

All of **The Five Cascades** are within a distance of an
eighth of a mile below Haines' Falls.   Crossing the creek at
the foot of the stairs, the top of the first Cascade is just in sight.
It is a fall of eighty feet.   The stream should be followed down,

on the right side, a little farther than this Cascade, where **Jacob's Ladder**—an extremely steep but substantial stair of some thirty steps—is to be descended. The first and second Cascades are near together, and the stream is recrossed between them. Another flight of stairs overhangs the side of the second Cascade, which is sixty feet high, and leads safely to its base. From here the path continues down the stream, on the left side, past the third and fourth Cascades, to

SHELVING ROCK AND NAIAD'S BATH.

The illustration here given of **Shelving Rock** conveys an excellent idea of its appearance. As may be seen, it projects far over the stream. The waters of the fifth Cascade tumble down underneath it, forming a beautiful pool, known as **Naiad's Bath,—**

> A lovely cave,
> Dusky and sacred to the Nymphs, whom men
> Call Naiads.—*Odissey.*

The cut shows the fourth Cascade, in the background. The combined picture of the Rock, Bath, Cascades, and the irimmediate surroundings is so enchanting that, when once seen, it is likely to be long retained in the memory.

A gentleman, who has traveled many years in California and Europe, pronounces the view from Shelving Rock the most charming one of the kind that he has ever beheld. People make repeated visits to this spot and prolong them for hours.

The ravine is walled in by perpendicular ledges the greater share of its length. On the north side, just below Shelving Rock, is a strata of rock, in some places five feet thick, containing considerable copper and sulphur.

SHELVING ROCK AND NAIAD'S BATH.

From Shelving Rock the walk may be continued down

THROUGH HAINES' RAVINE TO LAKE CREEK BRIDGE.

One hundred yards below Shelving Rock, by a path on the north side of the stream, will reach the foot of The Great Land-slide. If further procedure down the Ravine is undesirable, the Slide, itself, may be ascended to the road,—much more easily than its rugged appearance indicates. A hundred yards more, and the water pours over a perpendicular wall of solid rock to the depth of twenty-five feet. This is **Delmura Fall,**—the name being derived from the Spanish, meaning the wall.

The creek itself may now be followed and many places seen which will well repay the needed exertion ; or, the walk may be more easily accomplished by taking an old wood road close on the north bank, which leads to the Hunter Turnpike, at a point just above the Lake Creek Bridge and nearly opposite the path up the Kaaterskill Ravine.

HIGH PEAK.

WALK FROM THE HAINES' FALLS REGION.

From Haines' Falls it is four miles, in a southerly direction, to the summit of **High Peak.** The elevation of this mountain is 3664 feet, and it is conceded that its summit commands the most extensive view of any place in the Catskills. Parts of the path to it are not clearly marked, and it will be expedient to have a guide in taking this trip the first time.

The path crosses the Cauterskill just above Haines' Falls, and, after passing across a sloping field almost to the ledge of rocks at the upper side, turns to the left and enters the woods from near the south-eastern corner of the field. The ledge mentioned constitutes the Rifted Rocks, which are well worth visiting. (See description.) Before entering the woods, a glance to the east will reveal the Laurel House and Kaaterskill Falls, directly facing.

The route is now all the way through the dense forest. Following a wood road, of gradual ascent, along the side of the mountain for a mile, Santa Cruz Creek is reached. The beautiful Falls of the same name are but a little way down the stream and may be easily visited in this trip,—it being preferable to leave them for a stopping place in returning. (See description.)

Across Santa Cruz Creek the wood road turns to the right and becomes steeper. Other old roads diverge at various points, but the proper one to follow is most clearly defined. The first one to the left, and but a few rods from the Creek, is the path to Buttermilk, Wild Cat, and Viola Falls. (See description.) Two miles from Santa Cruz the road is left behind, and the journey is continued up the steep side of the Peak,—for a ways along or near the mossy bed of a tiny stream, —as indicated by the blazed trees. At the foot of an exceedingly steep pitch is **Comfort Spring,**—the source of the rivulet,—where it is customary to lunch, as water can scarcely ever be found nearer the top. Cole's last trip to the mountains was a visit to High Peak, and it was at this spring where he paused to take his noonday lunch.

Near the end of the ascent is a small, open space on a ledge of rock, which is called **Hurricane Ledge.** It affords an excellent view. To the north, over the long stretch of forest that has been passed through, the Cauterskill Clove lies some three thousand feet below. Up, beyond, the whole northern section of the mountain range may be seen, and between two of the peaks, Albany is discernible on a clear day. More to the east, quite a tract along and beyond the Hudson is visible.

The summit of High Peak is clothed with a growth of tall spruce and a thick carpet of green, velvety moss. The path does not lead directly to the pinnacle, but out along the eastern

side, on the verge of a precipice.   In the close foreground are
the Cauterskill and Plaaterkill Cloves ; then comes the Hudson
valley—the river visible north to Albany and far to the south ;
beyond, the Berkshire Hills, the Green Mountains, and other
parts of Massachusetts, Vermont, and parts of Connecticut ;
with the Adirondacks in the north—all together forming a natural panorama such as man seldom beholds.   The Upper Lake
by the Catskill Mountain House appears far below, like a tiny
mirror.

> There, as thou stand'st,
> The haunts of men below thee, and around
> The mountain summits, thy expanding heart
> Shall feel a kindred with that loftier world
> To which thou art translated, and partake
> The enlargement of thy vision.   Thou shalt look
> Upon the green and rolling forest tops,
> And down into the secrets of the glens,
> And streams, that with their bordering thickets strive
> To hide their windings.   Thou shalt gaze, at once,
> Here on white villages, and tilth, and herds,
> And swarming roads, and there on solitudes
> That only hear the torrent, and the wind,
> And eagle's shrieks. * * * * * *
> But the scene
> Is lovely round ; a beautiful river there
> Wanders amid the fresh and fertile meads,
> The paradise he made unto himself,
> Mining the soil for ages.   On each side
> The fields swell upward to the hills.—*Bryant.*

From a point a short way south of this outlook, and down a
ledge, there is a less obstructed and more extended view in
that direction.   The Overlook Mountain House is plainly visible on Overlook Mountain.

Returning to the previous point of observation, a path **may
be** found to the tip-top—a few yards distant.   Here, by **the**

help of a ladder to its lower branches, the tallest tree may be scaled by the venturesome. It is no easy feat; but one most estimable young lady of a party assured the writer that she had looked out upon the world from among its upper branches. It is quite an improvement on the views from below, as the other trees, in every direction, may be overlooked.

SANTA CRUZ FALLS.

WALK IN THE HAINES' FALLS REGION.

Santa Cruz Falls are one mile from Haines' Falls, on the way to High Peak. A description of the route to them may be found in the second and third paragraphs of the description of High Peak. (See index.) It is said that a bottle of Santa Cruz rum, supposed to have been left by some hunters, was found on the bank of the stream, from which circumstance this name was given. Having reached Santa Cruz Creek, it is but a few yards down the stream to the Falls. Either side of the Creek may be followed, but the best path is on the west bank.

The peculiar view disclosed, from the top of the Falls through the vista of the Cauterskill Clove, is remarkably fine.

> *   *   full many a spot
> Of hidden beauty have I chanced to espy
> Among the mountains ; never one like this ;—*Wordsworth.*

The descending spurs of the mountains dovetail into each other through the Clove ; beyond spreads the wide valley of the Hudson with the river stretching across it ; while the Berkshire Hills rise in the background,—the highest peak in the centre being Mt. Riga, near the junction of Massachusetts, Connecticut and New York. This picture is best seen in mid-afternoon,

as the sun is then in a position to properly light it up.   Durand
and other artists have spent much time at this spot.   There are
two Falls, but a stone's throw apart, each some sixty feet in
height.   At the upper one the descending water is much broken
by projecting points of rock ; but at the second, it pours down
in a broad, translucent sheet, and in the morning a rainbow is
frequently formed in the mist at the base,—so near that the
hand may be thrust among its colors.   It is this second Fall
which is represented in the illustration at the commencement of
the book.

The Falls may be descended on either side, with but comparatively little difficulty.   The ledge of rock which forms the
second Fall extends, in an arc, quite a distance to the right, so
that the shorter route is on the left side of the stream.   On
down the ravine are numerous cascades.

RIFTED ROCKS.

WALK IN THE HAINES' FALLS REGION.

The **Rifted Rocks** are across the Cauterskill, beyond
Haines' Falls, and in sight from the road.   The path is via the
top of the Falls—the same as in going to High Peak.

From a ledge of conglomerate, Nature has, at some period,
cut off a slice about twenty feet wide by three hundred feet
long and moved it out from the remaining part, so that there is
a passage behind, varying in width from ten to twenty feet

The strip is some fifteen feet high and broken crosswise into
blocks from twenty to forty feet long.   The entrance to the
passage is at the west end only, as, near the other extremity, a
sentinel of pudding-stone blocks the way.

nd
ıre in
en
wn is
he
all
of

m the
so
On

ond
the

iod,
feet
e is
feet
into
the
y, a

VIEW FROM

BUTTERMILK -FALLS, WILD CAT FALLS, VIOLA FALLS.

WALK FROM THE HAINES' FALLS REGION.

These three Falls are on separate streams that flow northerly from toward High Peak down in the Cauterskill Clove,—emptying into the Cauterskill Creek. Each one is about half a mile from the mouth of the stream on which it is located.

The first part of the path to them is the same as in going to High Peak, and the reader is referred to the article, "High Peak. Walk from the Haines' Falls Region." for a description. (See index.)

Having crossed Santa Cruz Creek and turned up the hill a few paces,—the path leaves the High Peak path, at nearly a right angle to the left, and continues easterly three-fourths of a mile to **Buttermilk Falls.** The first Fall is some seventy feet high, and a second one, just below, has about the same height.

Three-eighths of a mile farther along the path reaches the top of **Wild Cat Falls**—a sheer descent from a projecting ledge of about a hundred feet.

It is but a fourth of a mile more, by the same path, to **Viola Falls.** This takes its name from the fact that some violets were found in bloom near it as late in the season as the second of November. There are no very high Falls here, but the succession of small ones is so great that the water descends probably half a thousand feet in an eighth of a mile. Hillyer's Ravine is the name of the gorge through which the stream descends.

There are views from the tops of the Falls mentioned, down

into the Cauterskill Clove and out on the Hudson valley.   On **each** of the streams are other smaller falls besides the ones named.   The public have heretofore known but little concern. ing these Falls; but in a region less diversified by magnificent water-falls than that surrounding the Cauterskill Clove, these would have been sought out as attractions of a high order. The streams are, however, small, and perhaps people will not yet care to make this excursion, excepting just after rains have augmented the flow of water.

WALK FROM THE HAINES' FALLS REGION TO THE LAUREL HOUSE.

BY PATH VIA PROSPECT ROCK.

This is the shortest route and the one usually chosen to walk from one of these places to the other.   The distance is a mile and a half, and the path is shady the greater part of the way.

From the Haines' Falls House, the route is down the road past the Great Land-slide; up the rough road—Feather-bed Lane—to the north across a little bridge ; and up the steep hill in the field to the east.   At the top of the hill, near a lone hemlock tree, the path divides   (The lower one is the more direct way to reach the Sphinx.   For description, see index for "The Sphinx, or Noah's Ark.   Walk from the Haines' Falls Region."   The Laurel House may be reached by this way, but it is a trifle farther and the path is not quite as good.)   The upper one follows for a short distance, along a piece of woods ; then crosses a field ; passes through a grove and into a pasture. (In the edge of the pasture, on the south of the path, is a prom- inent bowlder, by which, a **path** leads down the hill to **The**

Sphinx.) Across the pasture the path again divides; the low-er one making a detour of a few steps to Prospect Rock and then uniting beyond with the upper one. (A description of the beauties seen from Prospect Rock may be found under the heading of "Prospect Rock. Walk from the Laurel House." See index.) On, a fourth of a mile from Prospect Rock, the path emerges from the woods and crosses a field—passing in its way by a Fish Pond—to the road, a short distance above the Laurel House.

This short walk gives a great diversity of scenery, and is a favorite one.

, THE SPHINX, OR NOAH'S ARK.

WALK FROM THE HAINES' FALLS REGION.

The Sphinx is on the brow of the mountain between Haines' Falls and Kaaterskill Falls, seven-eighths of a mile from the former place.

The route from the Haines' Falls House is down the road, a few yards past The Great Land-slide, to a rough road—Feather-bed Lane—on the left. Up the Lane a short distance a small bridge is crossed. Here the path leaves the Lane and ascends a steep hill in a partially cleared field on the right hand side. At the top of the hill, near an isolated hemlock tree, the path divides. (The upper one is the direct path to the Laurel House.) The one on the lower or right hand side is the most direct to the Sphinx, and, although part of it is not well defined, no one should have any difficulty in following it. Several trees and stones along the route are marked with spots of white paint. Beyond the hemlock tree the path—this part of it a wood road—soon enters the woods and at a fourth of a mile reaches a partially open space on the top of a ledge This is **Bellevue Point.** It overlooks the upper part of the Cauterskill Clove, with a glimpse of the Five Cascades to the west and Hunter and Rusk Mountains in the far background.

Continuing about a fourth of a mile farther, another open space is reached. Turning to the right, the path winds down a small ledge, easily passed, to The Sphinx. It is a curiously shaped rock, being about twenty feet high, with a base of ten feet square and its upper side projecting so as to resemble the prow of a ship. **Noah's Ark** was its original and obviously more appropriate appellation, although of late the other name has been more freely used. (There is another Sphinx on South Mountain.) The Ark stands close on the brink of a precipice on the side of the Cauterskill Clove, commanding a fine view thereof, and of the sides of High Peak and Round Top from base to summit.

The walk may be continued an eighth of a mile farther to

PROSPECT ROCK,

As follows : Proceeding northward from The Sphinx, through the clearing to the top of the hill, the upper path which diverges at the hemlock tree will be intersected. A few rods through the trees to the east is **Prospect Rock,** with its view so delightfully unique that none should fail to witness it. A description of this place may be found under the heading of "Prospect Rock. Walk from the Laurel House." (See index.) Prospect Rock may be conveniently visited from the direct path between the Haines' Falls and Laurel House Regions—it being but a few yards off that path.

WALK OR DRIVE FROM THE HAINES' FALLS RE-GION TO CLUM HILL.

Clum Hill is two miles and three-fourths from the Haines' Falls House, and by the same road for either a walk or drive from this place. The route is west, by the turnpike, a mile and a half, where a road branches to the south, in front of Maplewood. The remainder of the journey and the place are described under "Clum Hill. Drive in Tannersville." and "Clum Hill. Walk in Tannersville.' (See index.)

WALK FROM THE HAINES' FALLS REGION TO PARKER HILL.

The distance for a walk to **Parker Hill** is about three miles, and the route is as follows ·

One mile west from the Haines' Falls House the turnpike should be left, and the road to the right followed for some three-fourths of a mile, where another road intersects it on the west ;

Or,

There is a shorter way to reach this point of the journey By entering a path close by the west side of the Shady Grove House, a few rods will lead past the barns and through a clump of fir trees into the fields beyond. Across these fields, diag onally to the north-west, three-fourths of a mile will reach the road mentioned in the first route. By this road, the walk is continued toward the north-west ; passing two houses and, far ther on, two isolated barns A few yards beyond the second barn is the path up Parker Hill, on the right hand. The top of the Hill and its side toward the road were long since divested of their wood. For a description of the view obtained from this Hill. see "Parker Hill. Walk or drive from Tannersville." (See index.)

Parties taking this walk usually return by road by the way of Tannersville—a circuit of seven and a half miles.

It is not advisable to take the path mentioned via the Shady Grove House. immediately after a rain, as part of it is then quite muddy.

WALK FROM THE HAINES' FALLS REGION TO PLAATERKILL.

BY PATH AROUND THE SUMMIT OF HIGH PEAK.

This, the shortest route, is six miles ; by an old wood road, mostly through the forest. The first three miles is by the same path as to High Peak, and, therefore, the reader is referred to the article, "High Peak. Walk from the Haines' Falls Region," for a description of the same. (See index.) The remaining three miles of the old road is plain enough to be followed without difficulty, and leads to the head of the **Plaaterkill** Clove.

This route is, also, the shortest way from this place to the

OVERLOOK MOUNTAIN HOUSE,

And is continued from Plaaterkill,—making the total distance eleven miles.

LAUREL HOUSE REGION.

Besides the **Laurel House,** there are but two or three houses here. The Laurel House was named from the laurel—the *Kalmia latifolia* of botanists—which is profusely distributed through the forest immediately surrounding the house.

The location of the Laurel House is fifteen miles from Catskill, between the Catskill Mountain House and Haines' Falls —the last half mile leading thereto being a private road. Its elevation is 2038. feet. Standing just above the Kaaterskill Falls—which are Laurel House property—the view from the front of the house extends through the wide Kaaterskill Ravine, and embraces beyond, High Peak, Round Top, and other mountains, along the Schoharie valley.

A sea of fog sometimes fills the ravine below, but it is soon dispersed by the morning sun.

> Earth's children cleave to Earth—her frail
>   Decaying children dread decay.
> Yon wreath of mist that leaves the vale,
>   And lessens in the morning ray :
> Look how, by mountain rivulet,
>   It lingers as it upward creeps,
> And clings to fern and copsewood set
>   Along the green and dewey steeps ;
> Clings to the fragrant kalmia, clings
>   To precipices fringed with grass,
> Dark maples where the wood-thrush sings,
>   And bowers of fragrant sassafras.
> Yet all in vain—it passes still
>   From hold to hold, it cannot stay,

And in the very beams that fill
    The world with glory, wastes away,
Till, parting from the mountain's brow,
    It vanishes from human eye,
And that which sprung of earth is now
    A portion of the glorious sky.—*Bryant.*

From a point on the Laurel House road, but a few yards from where it leaves the Mountain House road, is a magnificent view of the mountains, from High Peak west to Hunter Mountain and beyond.

Conveyances from these houses run daily to Catskill to meet the boats and trains, and the route therefrom leads up the mountain through Sleepy Hollow. Carriages and stages can be obtained for drives to the many delightful resorts in the vicinity.

The **Mails** are received once or more each day—Catskill being the post-office.

A Western Union **Telegraph Office,** with direct wire **to** New York, is located in the Laurel House.

There is **trout** fishing in the neighborhood.

WALKS ABOUT THE LAUREL HOUSE REGION AND VICINITY.

DISTANCES FROM THE LAUREL HOUSE.

MILES.

Bastion Falls......... .... $\frac{1}{4}$
Catskill Mountain House, by path south of South Lake... $1\frac{1}{2}$
    "   "   by path via Druid Rocks....... $1\frac{1}{2}$
                  by path along north of South Lake $1\frac{3}{4}$
                  by road..................... $2\frac{1}{2}$
    "   by path via Sunset Rock....... 3

Council Bluff.. $\frac{3}{4}$

Fawn's Leap and Profile Rock, in Cauterskill Clove, via
Kaaterskill Ravine............................... 2

Glen Mary... $\frac{1}{8}$

Haines' Falls, by path via Prospect Rock.............. $1\frac{1}{2}$

" " by path via Kaaterskill Ravine and Haines'
Ravine,—including Bastion Falls, Shelving Rock and
The Five Cascades............................... $2\frac{1}{4}$

Haines' Falls, by road............................... $2\frac{1}{2}$

High Peak......... .... $5\frac{1}{2}$

Kaaterskill Falls.................................... o

Palenville, via Kaaterskill Ravine and Cauterskill Clove... 3

Palenville Overlook, by path via Lemon Squeezer, Fairy
Spring and Fat Man's Delight..................... $2\frac{3}{4}$

Prospect Rock...................... .... $\frac{1}{2}$

Santa Cruz Falls... .... $2\frac{1}{2}$

South Lake—Catskill Lakes....... .... $\frac{3}{4}$

Sunset Rock, South Mountain........................ I

The Outlook, on North Mountain.................... $2\frac{1}{2}$

The Sphinx, or Noah's Ark.. .... $\frac{3}{4}$

Descriptions of nearly all of the above walks are given in
these pages. (See index.) Some of the routes are so direct
that any description would be superfluous. The location of
places may be seen on the map.

DRIVES IN VICINITY OF THE LAUREL HOUSE.

DISTANCES FROM THE LAUREL HOUSE, BY THE MOST USUAL
ROUTES.

MILES.

Around South Mountain, via Mountain House, Sleepy
Hollow, Palenville, Cauterskill Clove, and Haines'
Falls,—round trip.................... 15
Catskill Mountain House........................ 2½
Grand View, East Windham....................... 21
Haines' Falls................................. 2½
Hunter 9
Overlook Mountain House, by new road via Plaaterkill.. 16½
Parker Hill.... .... 6½
Plaaterkill................................. 11½
Sleepy Hollow.............................. 4½
Stony Clove................................ 11
Tannersville 4½
Windham................................... 19

(See index to find description of above places.)

KAATERSKILL FALLS.

In the name **Kaaterskill Falls,** the original Dutch spell-
ing of the word has always been retained; but it is now more
frequently written Cauterskill, and, except when referring to
these Falls or the ravine leading therefrom, the latter form is
used throughout these pages.

It is but a few feet from the Laurel House to the top of **the**
Falls.   The **Spray House** stands on the very verge, and its
platform, with timbers bolted to the rock, projects over the aw-

KAATERSKILL FALLS.

ful chasm.   This is the point from which to view the Falls from
above; and over this first Fall the water drops a hundred and
sixty feet, broken into millions of foamy fragments ere it strikes
below, and flowing along a few yards it again plunges to the
depth of eighty feet.

> Midst greens and shades the Cauterskill leaps,
>   From cliffs where the wood-flower clings;
> All summer he moistens his verdant steps,
>   With the sweet light spray of the mountain springs;
> And he shakes the woods on the mountain side,
>   When they drip with the rains of autumn tide.
>
> But when in the forest bare and old,
>   The blast of December calls—
> He builds in the starlight, clear and cold,
>   A palace of ice where his torrent falls;
> With turret, and arch, and fretwork fair,
>   And pillars blue as the summer air. —*Bryant.*

The Kaaterskill Falls were the subject of one of Bryant's
beautiful poems,—the first portion of which is copied above,—
and in Cooper's "PIONEERS," Leather-Stocking says of them :—
"To my judgement, lad, it's the best piece of work that I've
met with in the woods;   *   *   *   *   I've sat on the shelving
rock many a long hour, boy, and watched the bubbles as they
shot by me, and thought how long it would be before that very
water which seemed made for the wilderness would be under
the bottom of a vessel, and tossing in the salt sea.   It is a spot
to make a man solemnize."

Gifford made a fine painting of the Ravine leading from the
Falls, and Durand, M'Entee, Cole, Church, and other well-
known artists have made visits to this delightful spot.

It is from under the Falls where its grandeur becomes most
striking.   At a gate by the Spray House a payment of twenty-

five cents is made—for once during the season—and a charming path followed a few yards through the forest to the head of
the stairs.   Rustic seats are placed along the way, and there
are resting-places at various landings along down the many
flights of stairs passed in reaching the bottom of the Falls.

In the immense amphitheater which curves behind the water
of the first Fall is a level path on which one may safely pass
entirely around behind the falling water.   Midway along the
path the flood comes pouring over the enormous arch of rock,
and as it descends, is eighty feet distant from the point of observation.   After passing around by this path, the stream may
be re-crossed a few yards below, at the top of the second Fall,
where the stairs continue down to the foot, and reach a seat
placed so as to give an unobstructed view of both Falls   While
parties are down here, the gate of a dam immediately above
the Falls is opened, thus augmenting the usual flow of water,
and the scene is then truly marvellous.

It is but a fourth of a mile down the stream to

BASTION FALLS.

The path to it crosses at the foot of the second Fall, and instead of following the creek, which has a sharp turn at this
point, it takes a more direct course along the wooded slope of
the ravine, meeting the creek just above **Bastion Falls** and
crossing it again by a foot-bridge of logs.

These Falls make a very pleasing picture, and a view of them
will convey the idea that the name is derived from the bastions
of rock which partially divide the stream as it breaks over **the
edge of the precipice.**

The path continues down

THROUGH KAATERSKILL RAVINE TO LAKE CREEK BRIDGE.

It is three fourths of a mile from Bastion Falls and the shaded path is all the way on the right side, quite close to the stream.

This path through the Kaaterskill Ravine—always a pleasant one—is much used as a short route for walks from the Laurel House to Fawn's Leap, Palenville, and up Haines' Ravine.

GLEN MARY.

WALK IN THE LAUREL HOUSE REGION.

Glen Mary is an eighth of a mile from the Laurel House, to the right of the Laundry. A rustic foot-bridge here crosses the Lake Creek, with its mossy rocks, just at the junction of Spruce Creek—the mingled waters passing over Kaaterskill Falls a few yards below.

SUNSET ROCK.—SOUTH MOUNTAIN.

WALK IN THE LAUREL HOUSE REGION.

Sunset Rock is a point on the side of South Mountain, overlooking the Cauterskill Clove. The distance from the Laurel House is one mile, by a good path through the forest. By the Laurel House Laundry the path crosses the creek to the right—below Glen Mary, which is visible,—and part of the way is a wood road. Half of a mile along, a path on the right leads to Council Bluff.

The view of the whole extent of Cauterskill Clove as disclos-ed from Sunset Rock is the best that can be found; and the vastness of High Peak,—seen from down in the Clove, where its sides are so irregularly cleft by the ravines leading from Buttermilk, Wild Cat, and Viola Falls, up over the dense forest to its summit,—is truly an imposing spectacle. To the west, Haines' Falls appear in the foreground, with the Schoharie valley and bordering mountains in the distance

There is a path over South Mountain from Sunset Rock to the Mountain House,—a two mile walk.

COUNCIL BLUFF.

WALK IN THE LAUREL HOUSE REGION.

Council Bluff is three-fourths of a mile from the Laurel House. The path to it is described under the heading of "Sunset Rock. Walk in the Laurel House Region." (See index.) The Bluff, which projects on the side of the Kaaterskill Ravine, affords a view down into the Cauterskill Clove and portions of Haines' Falls and Santa Cruz Falls.

WALK FROM THE LAUREL HOUSE TO THE HAINES' FALLS REGION.

BY PATH VIA PROSPECT ROCK.

This route is the shortest one between these places, and is much used for a walk. From the road, an eighth of a mile above the Laurel House, a path crosses the field to the west (passing in its way by the head of a **Fish Pond**—private property of the Laurel House), and enters the woods. A fourth of a mile through the woods it divides—the right hand branch continuing direct, while the left hand one leads a few steps out of the way to Prospect Rock, (for description, see index for "Prospect Rock. Walk from the Laurel House"), and unites with the other in a pasture a little beyond. (From this point a path turns down the hill on the left to The Sphinx, —see index for "The Sphinx, or Noah's Ark. Walk from the Haines' Falls Region,"—and this way may be used to reach Haines' Falls, but it is a little farther.) The direct path is readily followed the remainder of the way, and the whole walk is very enjoyable.

PROSPECT ROCK.

WALK FROM THE LAUREL HOUSE.

Prospect Rock is half a mile distant from the Laurel House by an extremely pleasing path. Up the road a little beyond the Laurel House barns the path enters the field to the left, through a gate ; crosses the upper end of the Fish Pond, and passes into the forest, through which it continues very clearly marked by being much trod. Care must be taken to turn to the left by the first diverging path, as the main one leads to the Haines' Falls Region, and Prospect Rock is a few steps off from this.

Prospect Rock is the best possible place from which to see the Kaaterskill Falls from a short distance It is about on a level with the top of the Falls, and a bold curve in the deep ravine places them directly facing. Both upper and lower Falls are visible from top to bottom and, also, the Laurel House, above.

> 'Twas here a youth of **dreamy mood,**
> A hundred winters ago—
> **Had wandered over the mighty wood,**

Where the panther's track was fresh on the snow ;
And keen were the winds that came to stir
The long dark boughs of the hemlock fir.

[*The Cauterskill Falls, by Bryant.*

It is but one-fourth of a mile farther to

THE SPHINX, OR NOAH'S ARK.

By a path, a little ways west from Prospect Rock, is a cleared space which slopes to the left and forms a V shaved piece ; and **The Sphinx** is at its lower extremity. A description of it is under the heading of "The Sphinx, or Noah's Ark. Walk from the Haines' Falls Region." (See index.)

WALK FROM THE LAUREL HOUSE TO HIGH PEAK.

The distance to **High Peak** by the shortest path is five miles and a half. A full description of the walk may be found under the headings of "Walk from the Laurel House to the Haines' Fall Region. By path via Prospect Rock," and "High Peak. Walk from the Haines' Falls Region." (See index). This trip may be made more easily by driving as far as Haines' Falls, thus reducing the walking distance to four miles. Along the route, a mile from Haines' Falls, the path is very near

SANTA CRUZ FALLS

And it will take but a few minutes to go to this charming spot. (See index for "Santa Cruz Falls. Walk in the Haines' Falls Region."

WALK FROM THE LAUREL HOUSE TO CATSKILL MOUNTAIN HOUSE.

BY PATH ALONG NORTH SHORE OF SOUTH LAKE.

The distance by this route is a mile and three-fourths, and the way is mostly shady   At the Laundry the path crosses the creek straight ahead, and beyond a few trees crosses a field to Scribner's road, near a small house.   Crossing the road. a half mile will reach the lower end of South Lake, from which point the path continues quite near to the Lake, on the north-west side, and strikes the Mountain House road near the Charcoal Pit.   The remainder of the walk is by the road.

BY PATH SOUTH OF SOUTH LAKE.

This walk of a mile and a half is the shortest route between these places, and is the one generally taken, although after rains it is somewhat muddy.   Most of the way is through the woods.   From the Laundry the path is to the right, across the foot-bridge at Glen Mary, to Scribner's.   Beyond the house, the left hand path should be taken.   It is clearly defined through the forest, and comes out by the Mountain House barns, passing Hygeia Spring on the way.

WALK FROM THE LAUREL HOUSE TO PALEN-VILLE OVERLOOK.

BY PATH VIA LEMON SQUEEZER, FAIRY SPRING, AND FAT MAN'S DELIGHT.

The shortest path from the Laurel House to **Palenville Overlook** is two miles and three-fourths.   To the right of the Laundry, the route is across the foot-bridge at Glen Mary, **to Scribner's.**   Beyond **the** house are three diverging **paths,** and the center one—an old wood road—should be followed. A mile through the woods it intersects a path from the Mountain House, near Druid Rocks. Turning to the right it ascends a small ledge and winds around to and underneath another ledge, soon passing the Lemon Sqeezer and Fairy Spring. (See index.) A few yards beyond, the path from Sunset Rock to the Mountain House is entered as it ascends the last ledge necessary to reach the summit of South Mountain. The path now crosses the narrow summit, passing by a bowlder known as Star Rock, to the front or eastern side of the mountain. Again turning to the right, a ledge is soon descended, to the left of which point is the U. S. Coast Survey Signal, and it is but a few rods more to the Bowlder. Passing down the ledge here, through the Fat Man's Delight (see index), the descending, tortuous path is quite well marked—in some places by piles of stones—for the remaining three-fourths of a mile to Palenville Overlook. (For description of this place, see index for " Palenville Overlook. Walk from Palenville."

CATSKILL MT. HOUSE REGION.

The **Catskill Mountain House** having existed for more than fifty years and being the largest house in the Catskills, it is quite generally known throughout this country as well as to many foreigners.   A mile and a half intervenes between this house and the nearest one to it.   It stands on the eastern side of the mountain range, on the edge of a high ledge of rock, overlooking the Hudson valley, and at an elevation of **2225** feet. It is plainly visible from a great distance to the east.

From the house a tract of many hundred square miles of the Hudson valley is spread to view ; with sixty miles of the river visible, appearing as a mere bright thread stretched across the patchwork plain of field and wood.   The Green Mountains and Berkshire Hills form the dim background, and the Highlands appear far to the right.

Directly south of the house, paths lead to **the** summit of South Mountain, which attains an altitude greater by **292** feet. The beautiful Catskill Lakes lie to the west a half mile and the road in that direction passes between them.

The route to the Mountain House is by the way of Catskill. **Stages** of the house meet the boats and cars at that place, which is twelve miles distant.   In ascending the mountain the road winds up through Sleepy Hollow ; and the rock **may** be seen where Rip Van Winkle is supposed to have slept after **quaffing** from "that wicked flagon !"   (For description of this place, see "Sleepy Hollow.") The stage fare from Catskill, including hand baggage, is two dollars. Trunks are charged for extra.

Conveyances may be had for drives, the principal ones being to Kaaterskill Falls, Haines' Falls, Stony Clove, and around South Mountain through the Cauterskill Clove.

The **Mail** is carried regularly by the stages of the house to and from the Catskill post-office.

A Western Union **Telegraph Office** is in the house with a direct wire to New York.

WALKS ABOUT THE CATSKILL MOUNTAIN HOUSE REGION AND VICINITY.

DISTANCES FROM THE MOUNTAIN HOUSE.

MILES.

Artist's Rock, and Prospect Ledge........................ $\frac{1}{2}$

Catskill Lakes, (North Lake, South Lake), by road....... $\frac{1}{2}$

Haines' Falls, by path via Laurel House................. 3

    "   " by road.. $4\frac{3}{8}$

Hygeia Spring..

Kaaterskill Falls, by path south of South Lake........... $1\frac{1}{2}$

    "   " by path along north shore of South Lake. $1\frac{3}{4}$

    "   " by road............................ $2\frac{1}{2}$

Lovers' Retreat.. $\frac{1}{8}$

Mary's Glen... 1

Moses' Rock.. $1\frac{1}{4}$

Newman's Ledge .... $1\frac{3}{4}$

Palenville Overlook, by path over South Mt., via Pudding-

    stone Hall and Fat Man's Delight.... .... $1\frac{1}{4}$

South Mountain,—top of it, by direct path............ $\frac{1}{2}$

"   "   Circuit of the top, by path via Druid Rocks,
Lemon Squeezer, Fairy Spring, and Pudding-stone Hall $1\frac{1}{4}$

Sunset Rock, on South Mt., by path over the Mt......... 2

"   "   "   "   via Scribner's.............. $2\frac{1}{2}$

Sunset Rock, north of Mt. House, ⎫

Jacob's Ladder,   ⎬ .... $\frac{7}{8}$

Bears' Den,   ⎭

The Outlook, on North Mountain..................... $2\frac{3}{4}$

DRIVES IN THE VICINITY OF THE CATSKILL MOUNTAIN HOUSE.

DISTANCES FROM THE MT. HOUSE, BY THE MOST USUAL ROUTE.

MILES.

Around South Mountain, via Sleepy Hollow, Palenville,
Cauterskill Clove, and Haines' Falls,—round trip,.... 15

Haines' Falls 4

Hunter 10

Kaaterskill Falls and Laurel House .... $2\frac{1}{2}$

Parker Hill.. 8

Plaaterkill, ... 13

Sleepy Hollow..................................... 2

Stony Clove....................................... $12\frac{1}{2}$

Tannersville 6

SOUTH MOUNTAIN.

It is half a mile from the Mountain House to the summit of **South Mountain** by the shortest path.   This walk is probably more frequently taken than any other here.

The path ascends among the trees from near the south-west corner of the Mountain House.   A painted guide-mark along it consists of a round, white spot and a red bar crossing it. Passing up a small ledge and reaching the top of a second, higher one, paths diverge.   The one to the summit is through **Pudding-stone Hall** and up a third ledge.   This Hall is a narrow passage between the ledge and a huge block of pudding-stone broken and separated therefrom.   The plain path winds along the eastern ledge of the summit and commands views much the same as those from the Mountain House but much increased in extent and variety.

> I looked upon a plain of green,
>   That some one called the Land of Prose,
> Where many living things are seen,
>   In movement or repose.
>
> *   *   *   *   *   *   *   *   *
>
> But most this fact my wonder bred,
>   Though known by all the nobly wise,
> It was the mountain streams that fed
>   The fair green plain's amenities —*Stirling.*

The summit of South Mountain and much of its sides are nearly barren.   Slightly back from the eastern edge is a bowlder on the very top, called **Star Rock.**   The elevation here is 2,497 feet.   At the edge of the mountain, east from this point, a part of the projecting ledge is called the Sphinx.

The path continues southerly to Palenville Overlook, which is a mile and a fourth from the Mountain House. Descending one ledge, a signal used by the U. S. Coast Survey stands on the left. A few yards further along the **Bowlder** appears on the edge of a high ledge. To the left of the Bowlder is the only near, convenient way down. A rent in the ledge, called the **Fat Man's Delight**, affords passage through, although the fat man may find difficulty in getting through the thirty feet of its length. From here the path is quite winding down through the shrubbery to Palenville Overlook. (For description, see index for "Palenville Overlook. Walk from Palenville."

CIRCUIT OF THE TOP.

Near the south-west corner of the Mountain House the same path is taken as in going to the summit of the mountain. At the top of the second ledge it separates to the right and winds along below the third ledge, through the forest, nearly half a mile to **Druid Rocks.** These Rocks are some blocks of conglomerate, on the left of the path ; one piece, about a dozen feet high, being called the **Great Bear,** as it somewhat resembles a large animal in a sitting posture. On, a ways farther the path is near the ledge, but then it turns and ascends it, and by quite a circuit to the left reaches a ledge above and turns to the right underneath it, through a fissure— the **Lemon Squeezer,** or, the **Elfin Pass**—passing **Fairy Spring.** It soon ascends the ledge to Star Rock. The return is made by way of Pudding-stone Hall, the same as in the walk by direct path to the top.

This makes a walk of about a mile and a fourth.

CATSKILL LAKES.

WALK IN THE MT. HOUSE REGION

The **Catskill Lakes** are a half mile west of the Mountain House. They are distinguished as North and South Lake, and lie close on either side of the road to Kaaterskill Falls and Haines' Falls. The North one has an area of about twenty acres, and the South one some thirty-five acres The water from them flows to the south-west over the Kaaterskill Falls. Their elevation is 2,138 feet. They are bordered by forest—one side of the South one by the side of South Mountain—and the Shore of the North one is covered with a cranberry marsh. Before reaching the Lakes by way of the road, a path turns off to the left and strikes the shore of South Lake at the boat-house.

VIEW ON SOUTH LAKE

Till death the tide of thought may stem,
There's little chance of our forgetting
The highland lake, the water gem,
With all its rugged mountain-setting.—*Milnes.*

A few yards east of the Lakes, close on the south side of the road, is a large rock with a crevice in one end resembling **the** open jaws of an Alligator, and hence called **Alligator Rock.**

ARTIST S ROCK, AND PROSPECT LEDGE.

WALK IN THE MT. HOUSE REGION.

It is a walk of half a mile from the Mountain House to **Artist's Rock,** which is a bowlder on the brow of the mountain north of the house. The way is down the mountain road to the top of the second small hill. where two paths start on the left. The right hand path gradually ascends a ledge, giving a sight of the valley below. and soon reaches Artist's Rock from where the scenic beauty is seen with increased extent.

After ascending some steps in the rocky way it is only a few yards to **Prospect Ledge.** At this place the high wall of rock facing the valley curves farther out than at any other point near, and thus greatly extends the view both north and south. The Hudson valley is spread to view form Albany to West Point. Down to the left and near is Sleepy Hollow, while just over it is Cairo Round Top—a prominent round hill. This is an attractive walk.

On, three-eighths of a mile are

SUNSET ROCK, JACOB'S LADDER, AND BEARS' DEN

From Prospect Ledge the path soon ascends a short hill,—called Red Hill, from the color of the earth,—at the top of which stands a bowlder on the right of the path, known as **Sunset Rock.** The views from this point are Interesting Stone steps and a ladder—**Jacob's Ladder**—permit of the next ledge being ascended, and to the left of the top are deep clefts in the rock, constituting the **Bears' Den.** To the south appear the Mountain House resting against South Mountain, with High Peak and Round Top looming up in the distance. Close down to the west lie the Catskill Lakes and a dozen miles beyond appear Stony Mountain, Hunter Mount-ain, and Rusk Mountain. This spot commands a varied scenery that is well worth viewing.

The path continues to

NEWMAN'S LEDGE,

Which is a mile and three-fourths from the Mountain House. After winding along a ways it ascends quite a steep hill, and near the top of this hill it turns to the right to **Newman's Ledge.** (The path which continues up the hill leads to The Outlook.) The scenes presented from this Ledge are similar to those seen from Prospect Ledge except that the northern horizon has receded and the southern drawn nearer. The Ledge was named after Rev. Newman Hall. It has a per-pendicular height of about one hundred feet.

THE OUTLOOK, ON NORTH MT.

The Outlook is two miles and three-fourths from the Mountain House. Following the path from where the one branches to Newman's Ledge beyond the Bears' Den, a small hollow is crossed in which, on the left, is a depression called **The Cellar.** The way is now mostly through an unbroken forest ; at one place through a clump of tall, straight spruce up the slope of the mountain, with the ground exceedingly smooth and free from brush—as beautiful a piece of wood as is often met with. 'The path through the forest is clearly shown by marked trees.

There are two points for observation from The Outlook, called the First and Second Outlook. This brow of North Mountain on which they are situated is extremely steep. From the First a broad. steep basin lies below and in its depths are the Catskill Lakes. In successive order, beyond are the Mountain House, South Mountain, and High Peak—the latter seen just to the left of the curve in the mountain forming the other Outlook Round Top is hidden by this curve. From here a good idea may be obtained of the route and distance traveled to reach this point.

The Second Outlook is a few rods farther along the path. Its elevation is 3,108 feet. The view is similar to that obtained from the First, with some portions increased in beauty and extent.

The return route is usually by a nearly direct and plain path from the Second Outlook down through the forest, striking the road just west of the Lakes

WALK FROM THE CATSKILL MT. HOUSE TO KAATERSKILL FALLS.

BY PATH SOUTH OF SOUTH LAKE.

The distance is a mile and a half. The path starts by the barns, passes Hygeia Spring three-eighths of a mile along and continues most of the way through the woods to Scribner's. Here it passes front of the house, and beyond the saw-mill crosses the creek at Glen Mary and comes out by the Laurel House Laundry. This is the shortest path between these places and gives a charming walk ; but some parts of the way are usually muddy just after rains.

BY PATH ALONG NORTH SHORE OF SOUTH LAKE.

This route is a mile and three-fourths long. From on the road, west of the Lakes, near the charcoal pit, a well worn path follows through the woods. near the Lake, and comes to the road to Scribner's, and opposite this point is a small house. The path continues close by this house, on the right, across a field and over Spruce Creek by the Laundry.

SLEEPY HOLLOW.

The spot known as **Sleepy Hollow,** and said to be the place where Rip Van Winkle took his twenty years' nap, is on the Mountain House road two miles down the mountain.

The illustration showing Rip at home is from the piece of statuary by John Rogers. (For list of other groups, see index for "Rogers' Statuary.") Irving says:

"The children of the village, too, would shout with joy whenever he approached. Whenever he went dodging about the village, he was surrounded by a troop of them, hanging on his skirts, clambering on his back, and playing a thousand tricks on him with impunity; and not a dog would bark at him throughout the neighborhood."

MARY'S GLEN.

WALK IN THE MT. HOUSE REGION.

The shady walk of a mile, without climbing, from the Mountain House to **Mary's Glen** is a desirable one. The way is down the mountain road to the top of the second small hill, where the left one of the two paths on the north should be followed. It leads past the eastern end of North Lake, crossing a small stream near the Lake. Half a mile farther Ashley's Creek is crossed on a log for a foot-bridge. A pretty falls are some two hundred feet further up the stream. A path leads from the top of the falls to the road, by the charcoal pit west of the lake, and the return is usually by this way

HYGEIA SPRING.

WALK IN THE MT. HOUSE REGION.

Hygeia Spring is three-eighths of a mile from the Mountain House, on the path which passes east of the barns and enters the woods. It is a nice spring of cold water, walled up, and with conveniences for drinking.

MOSES' ROCK.

WALK IN THE MT. HOUSE REGION.

Moses' Rock is on the eastern side of the mountain, below the Mountain House. The way to it is down the mountain road about half a mile and then by a path on the right down the mountain three-fourths of a mile farther. (From the path a short distance from the road a path diverges and leads along the side of the mountain to Palenville Overlook.) The Rock is about one hundred feet long by twenty feet high, and water flows from an opening in its side a yard above the ground.

(This path continues down the mountain and is the shortest way to Palenville.)

TANNERSVILLE.

About half a century ago the name of **Tannersville** was borne by a little hamlet with its post office at the foot of the mountain, in the upper part of the Cauterskill Clove, which owed its existence chiefly, if not wholly, to a large and flourishing tannery located there   In later years, as the tannery ceased operations and the place became one of ruins, the name was transferred to the locality which has ever since retained it, and which is three miles and a half from the original site.

Tannersville is the next place west of the Haines Falls Region, on the same road,—the Hunter turnpike,—and is fifteen miles from Catskill. Its nearness to the Catskill Mountain House, Kaaterskill Falls, Haines' Falls, Cauterskill Clove and Stony Clove renders excursions to these places easy to be made; while many other interesting points, but little more remote, may be conveniently visited.

Tannersville has quite a number of popular boarding-houses, with accommodations for from fifteen to a hundred people each; and as the number of people' who stop at this place increases each succeeding year, additional provision is made for their reception.

The general elevation of the place, where the houses are located, varies from one thousand and eight hundred to two thousand and two hundred feet. Clum Hill is in the immediate south, and Parker Mountain and Parker Hill rise on the north

side.   To the west is a grand view of Hunter Mountain with
its many irregular spurs, and the view down the Schoharie Valley is, also, an excellent one.

The route from Catskill to Tannersville leads through the
well-known Cauterskill Clove, up the mountain.   Nearly all of
the houses in Tannersville run daily **conveyances** to meet
the boats and cars at Catskill; and, by previous arrangement,
parties will be met at unusual hours.   The regular fare, including baggage, from Catskill is two dollars.   Besides the conveyances from various houses, there is, also, a daily line of mail
stages between this place and Catskill.

There are plenty of vehicles obtainable for local drives.

Mails are received from New York once or more each day.
The post-ofhce is at the Mountain Home.

The **Telegraph Office** is at the Mountain Home—on the
Hunter, Tannersville and East Jewett line.

A Union **Church** has recently been erected.   There is a
store in the place.

Near the Mountain Home is a turning shop where many
fancy articles of wood are turned, in a style seldom equaled
and probably never excelled.   No lover of pleasing designs and
artistic workmanship can fail to appreciate a visit to this shop.

The piscatorial art may be practiced in the vicinity, and the
streams hereabout are being re-stocked with **trout** from the
county hatchery.

WALKS ABOUT TANNERSVILLE AND VICINITY.

DISTANCES FROM THE MOUNTAIN HOME.

MILES.

Clum Hill... 1

Haines' Falls 2½

Haines' Falls Ravine, as far as { The Five Cascades, Shelving Rock and Naiad's Bath, } .. 2¾

Haines' Falls Ravine,—through it to Lake Creek Bridge 3⅝

Kaaterskill Falls and Laurel House, by path via Prospect Rock.. 4

Kaaterskill Falls and Laurel House, by road.......... 4½

Lovers' Retreat...................................... 1-16

Old Indian Fort, between High Peak and Round Top.. 7

Parker Hill.. 2

Raspberry Lane.......... .... ½

Nearly all of the above walks are described in these pages. (See index). Where the route is by a direct road, the description is omitted. The location of these places and ways to them may be seen on the map.

DRIVES IN THE VICINITY OF TANNERSVILLE.

DISTANCES FROM THE MOUNTAIN HOME, BY THE MOST USUAL ROUTES.

MILES.

Around South Mountain, via Mountain House, Palenville, and Cauterskill Clove,—round trip............... 18½

Catskill Mountain House........................... 6

Cauterskill Clove,—through it to Palenville........... 6

Clum Hill 8

Haines' Falls,.. $2\frac{1}{2}$

Hunter, ... $4\frac{1}{2}$

Kaaterskill Falls and Laurel House,................... $4\frac{1}{2}$

Overlook Mountain House, by new road, via Plaaterkill,.. 12

Parker Hill,.. 2

Plaaterkill, ... 7

Sleepy Hollow, where Rip Van Winkle slept!........... $8\frac{3}{4}$

Stony Clove,.. $6\frac{1}{2}$

Windham, ... $14\frac{1}{2}$

(See index to find description of above places).

CLUM HILL.

WALK IN TANNERSVILLE.

The walk to **Clum Hill** is considered the most important and most popular one in Tannersville. The Hill is a continuation of the ridge formed by High Peak and Round Top, and was named after its owner. Its elevation is 2372 feet. It rises on the south of Tannersville and the distance from the Mountain Home to its summit, by the shortest path, is one mile.

By the road from directly in front of the Mountain Home to the south a fourth of a mile, will cross two bridges. Just beyond the second bridge the path starts from the road, crosses the fields to the foot of the Hill, and winds up its north side, which is covered with a growth of shrubs.

The view from the summit is panoramic. Looking over Tannersville hamlet, to the north, shows Eastkill and Parker Mountains and Parker Hill, while in the background are five other prominent peaks. The view to the east embraces the Laurel House and Kaaterskill Falls, which are some three **and** a half miles distant in a direct line   The Falls are facing this point and show the falling water from top to bottom.   Round Top is in the near southeast.   To its right lies the head of the Plaaterkill Clove, and beyond rises Plaaterkill Mountain.   From Plaaterkill Mountain the eye may follow along the ridge to the west, passing successively: Indian Head, Schoharie Peak, Mink Mountain, Stony Mountain, Hunter Mountain, Colonel's Chair, Rusk Mountain, and others on down the valley, becoming less and less distinct, till they fade away into the far blue horizon.   There is no better view obtainable of this range of mountains and of the Schoharie valley than the one from Clum Hill.   Hunter Village lies just behind a long-extended spur of Eastkill Mountain.

(There is a road nearly to the top of this Hill.   See index for "Clum Hill.   Drive in Tannersville.")

RASPBERRY LANE.

WALK IN TANNERSVILLE.

In front of Elm Cottage, and one-half of a mile west of the Mountain Home, **Raspberry Lane** extends from the turnpike, across to the road to Plaaterkill—entering it near Blythewood—and is about three fourths of a mile long.

Midway along the Lane is a grove of hemlocks and maples, in which picnics are frequently held.   The pretty brook which flows through among the trees, renders the spot more inviting, and the walk here is a pleasant one at all times. The Schoharie creek is a few rods below, and is crossed if the walk is continued to the Plaaterkill road.

LOVERS' RETREAT.

WALK IN TANNERSVILLE.

Lovers' Retreat is among the cluster of trees, down the
road a few rods, in front of the Mountain Home.   Seats are
arranged in this secluded shade, on a bank at whose foot winds
a pebbly trout brook.

> Beneath the quivering arch of leaves
>   Where sunlight flickered through,
> While birds sang merry songs of love,
>   Each to its mate so true,
> Where just below the mossy bank
>   The laughing stream flowed by,
> We came, with fishing line and rod,
>   My blue-eyed May and I.—*Harpers' Magazine.*

PARKER HILL.

WALK OR DRIVE IN TANNERSVILLE.

The proper route to **Parker Hill** is by the direct road at
the west side of the Mountain Home.   It leads north, with
gradual ascent, for nearly two miles, where it is intersected by
a road on the right.   A few rods along this right hand road is a
path to the left, which leads up through the field to the summit
of the Hill.   It is but a few rods from the road to the top.

The elevation of this mountain is 2545 feet.   The north side
of it is quite precipitous, and thus allows an excellent sight
down through the Eastkill valley,—which descends to the west,
—taking in East Jewett.   Beyond, is a tract of hilly country
and numerous peaks in the direction of Hensonville and Windham.   More to the right, and just across the Eastkill valley,

grandly rise several high mountains—Black Head farthest to the east and Black Dome next, adjoining. There is no other walk or drive in this vicinity that gives such an idea of the Eastkill valley and contiguous country as this view from Parker Hill.

Prominent in the south-east are High Peak and Round Top, and the view of them from here is an impressive one. To the west of Round Top the successive peaks can be well seen as far along as Hunter Mountain; while Tannersville lies in the foreground.

The mountain immediately west of Parker Hill is Parker Mountain.

In ascending Parker Hill a boulder is passed, on which a large white star has been painted. For this reason the original name—Parker Hill—has, for a time, been lost sight of, and the place been known as **Star Rock.** The boulder is not where the view is obtained, nor in any way specially connected therewith; therefore the first appellation is retained in these pages.

The trip to this Hill is a popular one.

WALK FROM TANNERSVILLE TO THE LAUREL HOUSE.

BY PATH VIA PROSPECT ROCK.

The first part of this walk is by the turnpike—two and a half miles. A detailed description of the remainder of the route may be found under the heading of "Walk between the Haines' Falls Region and the Laurel House. By path via Prospect Rock." (See index.) The total distance from the Mountain Home, by this route, is four miles.

CLUM HILL.

DRIVE IN TANNERSVILLE.

It is practicable to drive almost to the summit of **Clum Hill.** By the way of this road the distance is two miles from the Mountain Home. The turnpike is followed east for a mile where a road to the south is taken,—directly in front of Maplewood,—which descends a ways and then winds up the long, steep hill to the top of the ridge, terminating in an old road running perpendicular to it. This old road may be followed a few rods past the house to the west, and then the path crosses a field to the summit.

An enumeration of the beauties seen from this place may be found under the heading of "Clum Hill. Walk in Tannersville." (See index).

HUNTER.

The village of **Hunter** is located on the Schoharie Creek at the base of Hunter Mountain, and forms a part of the township bearing the same name. The township extends from west of the village, east to Palenville, and has an area of about one hundred square miles. The place was named after John Hunter, who owned a large tract of the mountains here during its early settlement. In the year 1717, Colonel William W. Edwards and his son moved from Massachusetts to this Village and here established the first extensive tannery in the State, in which the then new method of tanning was adopted. During the Revolutionary War the Indians often crossed the mountains by way of the Schoharie valley, and tories in this region acted as guides in some of their plundering expeditions.

Nearly every building in the village of Hunter is on the one street which follows the course of the Schoharie for a mile and a half. The greater portion of the street is shaded,—for a ways by a row of lordly old elms, and at other places by maples. The elevation is from sixteen to seventeen hundred feet

Hunter is west of Tannersville and adjoining, and is twenty miles from Catskill. The route to it is usually by the way of Phœnicia, on the Ulster and Delaware Railroad, thirteen miles distant. **Stages** meet the trains, and private conveyances are sent for parties when such arrangements are previously made. **The road** leads through the remarkable Stony Clove, where ice remains among the rocks during the entire year. (For description of this place, see index for "Stony Clove.") The stage fare from Phœnicia, with baggage included, is one dollar.

Conveyances may be had for drives to Catskill Mountain House, Overlook Mountain House, Kaaterskill Falls, Haines' Falls, and many other popular summer resorts, more or less distant.

There is a daily **Mail** between Hunter and New York. The post-office is at the store of Burgess and Douglass.

There are two **Telegraph Offices ;** one at the upper end of the village on the Hunter, Tannersville and East Jewett line, and the other near the lower end, at the store of H. E. Biddell & Co., on the U. & D. Railroad Company's line.

The three **Churches** are of the Methodist Episcopal, Presbyterian, and Roman Catholic denominations.

There are two physicians in the place ; three stores ; two extensive chair factories, which are continually producing various styles of chairs in great numbers ; and a bedstead factory.

WALKS ABOUT HUNTER AND VICINITY.

DISTANCES FROM THE POST-OFFICE.

	MILES.
Chair Factories, in the village....	$\frac{3}{4}$ and $1\frac{1}{2}$
Cold Spring, in Shanty Hollow.................... ..	1
Colonel's Chair......................................	2
Entrance to Stony Clove Notch......................	3
Ford Hill...	2
Hunter Mountain.......,...........................	4
Ingraham Square,—around it........................	5
Lovers' Glen...	$1\frac{1}{4}$

CENTRAL HOUSE, HUNTER.

Mossy Brook..   1

Rusk's Hill.. $\frac{1}{2}$

DRIVES IN THE VICINITY OF HUNTER.

DISTANCES FROM THE POST-OFFICE, BY THE MOST USUAL ROUTES.

	MILES.
Càtskill Mountain House..........................	10
Clum Hill...	6½
Grand View, East Windham........................	12
Haines' Falls......................................	7
Kaaterskill Falls and Laurel House.................	9
Overlook Mountain House, by new road via Plaaterkill..	14
Parker Hill.......................................	7
Plaaterkill.......................................	9
Prattsville	16
Stony Clove	5½
Tannersville.......................................	4
Windham	10

(See index to find description of above places.)

HUNTER MOUNTAIN.

WALK FROM HUNTER.

Hunter Mountain is south of Hunter village, and rises abruptly, with many irregular spurs, from the Schoharie valley. It has an elevation of 4040 feet, and is the highest peak of the Catskills. The summit is 2431 feet above the village of Hunter, and the distance thereto by the shortest path is about four miles, although as measured on a map it would appear to be but one-half as far.

Of the several different routes by which the writer has made the ascent of this mountain, it may be said that none have been as satisfactory, in point of freedom from difficulties and economy of time, as one taken by an almost direct course from the village up its steep side.

By this way, a path crosses the Schoharie in front of the post-office and leads up through some fields and a strip of woods to Shanty Hollow, a mile distant, where there is a small farm house. The upper part of the Hollow is divided by a long spur of the mountain, and the way continues, east of the house, across a small stream, by a log road up the eastern branch of the Hollow for three-fourths of a mile. From here the way is exceedingly steep and there is no path marked. Crossing to the west side of the rill which comes down this hollow, a proper course to follow by the compass is about forty degrees west of south. It will be well to bear in mind that there are no more hollows to cross in any part of the journey, and that, therefore, the route is always ascending. There are no high ledges to form barriers to one's progress, but the steep mile, of quite regular slope, rises at some places at an angle of forty-five degrees or more.

From the top of this steep portion, the remainder of the journey is of gradual ascent, and should be continued in the same direction. The long walk through the dense forest here is delightful.

The summit of Hunter Mountain is a level area of about one-fourth of a square mile, covered with a thick growth of spruce, as are most of the other peaks of the Catskills. Owing to the forest, the views obtained from here are quite meager;

however, from a small ledge on the eastern brow, High Peak and Round Top may be seen, seven miles distant, and from the west side it is possible to get a glimpse down into Westkill. Bears and deer have well-worn paths converging at a small depression, near the eastern outlook, which holds water for some days after rains. The south side of the mountain descends by many ledges towards Stony Clove; while a mile and a half, along the high ridge which extends south of east, the deep and narrow Stony Clove Notch cuts through and separates this mountain from Stony Mountain. The adjoining peaks, continuing in this direction from Stony Mountain, are Mink Mountain, Schoharie Peak Indian Head, and Plaaterkill Mountain, which latter terminates this ridge at the front or eastern side of the mountain range. By a branch of Hunter Mountain to the west, the first peak encountered is Westkill Mountain, while others follow in the same direction. By another branch, more to the north, Rusk Mountain is first met, beyond which is a succession of peaks down the Schoharie valley. West of north a long, curved spur forms the Colonel's Chair.

Quite a large number of people ascend Hunter Mountain each summer. Probably ladies will not care to make the attempt. The contour of this mountain is such that, without some use of the compass, it will be a very easy matter to descend in an unintended direction.

COLONEL'S CHAIR.

WALK FROM HUNTER.

The **Colonel's Chair** is a long spur of Hunter Mountain which extends towards the village and terminates on the Schoharie.

In front of the post-office, the path to it crosses the creek and the fields to Shanty Hollow, a mile distant. The stream along the way, on the left, is **Mossy Brook,** and is frequently visited. In Shanty Hollow, the way is to the right of the little house, and the point at which the shrubbery beyond should be entered is not clearly indicated, but once properly started, the remainder of the way may be readily found.

A few rods back of the house in the Hollow is **Cold Spring,** where enough clear, cold water issues from the ground to supply the village, if pipes were laid to conduct it thereto.

The ascent of the last half mile of the Chair, over the broken rocks with which its sides are covered, must be considered as climbing rather than walking, for it is so steep that the hands will frequently have to assist locomotion.

The ridge forming the Chair measures but a few rods across its summit. Its elevation at the front is 3037 feet, and that of the highest place, a fourth of a mile back, is 3165.

The summit is almost barren, and the view from here is an extensive one, embracing many miles along the Schoharie valley and the surrounding country.

Many ladies make this trip and feel well repaid for the needed effort.

STONY CLOVE.

It is **Stony Clove Notch** to which attention is especially given under this heading. Its remarkable features are unlike those of any other place in the Catskills. Nature has here cut a deep, narrow pass through the loftiest ridge. The entrance to this Notch is three milles from Hunter village and four miles from Tannersville. The road which passes through extends to Phœnicia.

STONY CLOVE NOTCH.

Proceeding through from the north, or Hunter side, the gap gradually becomes narrower, with its sides considerably steeper and higher, and the road somewhat ascending, for a mile or more, when the highest point on the road and the narrowest part of the Notch are reached. Here the sides rise almost perpendicularly, more than two thousand feet, and are so close that the roadway will not admit of wagons passing each other except in a few places. There is no creek running through the Notch, but the tiny rills which come down the sides pass out, either toward the north to the Schoharie, or else southward to the Esopus,—taking circuitous and widely separated courses to the Hudson. As the sun never shines in some portions of the Notch, ice, which is formed among the rocks, remains during the entire year, and is obtained from cavities close by the road, even in mid-summer.

Continuing through, the road descends gradually. The pass widens, and the right hand side is a high wall of rock, in some places overhanging. At its base, near the outlet, is a pool of dark water, called **Stygian Lake.** Farther along, close to the other side of the road, is the **Devil's Tombstone,**—a bowlder planted endwise in the ground and rising a dozen feet above the surface. This is also known as **Pulpit Rock,** and, as parties frequently picnic at this spot, **Picnic Rock** is another appellation. It is possible to obtain trout from the stream near by, which flows from Stygian Lake.

It is about half a mile from here to the inhabited part of Stony Clove,—which contains a few houses, saw-mills, and a chair factory, nestled between the high mountains on either side,—through which the road takes a westerly direction along the winding course of the Stony Clove Creek. There are many trout in this stream.

OVERLOOK MT. HOUSE REGION.

The **Overlook Mountain House** is located near the summit of Overlook Mountain, at an elevation of 2978 feet— higher than any other house in the Catskills.

The route to this house from the Hudson is nine miles by the Ulster and Delaware Railroad from Rondout to West Hurley ; thence nine miles by **stages** of the Overlook House which meet all trains.   The fare on the cars is thirty-one cents. The stage fare is one dollar and a half, exclusive of baggage.

The Overlook Mountain stands at the south-eastern corner of the Catskill range and reaches an altitude of 3150 feet. The mountain and also the house may be observed for a long distance along the Hudson.

The mountain is nearly north from West Hurley and the view of it from that point, as it towers above, with the house in full view, is exceedingly grand.   Four miles from West Hurley the stages come to the little village of Woodstock which has an elevation of 594 feet.   This is the **Post-office** of the Overlook House and the stages carry the **Mails** to and fro twice or more each day.

From Woodstock the ascent of the mountain is made by a good road through the forest.   Half way up the mountain is the nearest house to the Overlook House.

The house, standing close on the brow of the mountain, commands surprising and unbroken views in nearly every direction. Down in the front is a plain of ten miles reaching to the Hudson, with a long stretch of the river and contiguous country.

OVERLOOK MOUNTAIN AND OVERLOOK MOUNTAIN HOUSE.

To the south and west appear Slide Mountain and adjacent mountains.   Portions of seven states are said to be visible from here.

There is a **Telegraph Office** in the house.

Conveyances may be had for **drives** to Cauterskill Clove, Haines' Falls, Kaaterskill Falls, and Catskill Mountain House.

WALK ABOUT THE SUMMIT OF OVERLOOK MT.

Standing just in front of the Overlook House and facing the valley, to the left the front of the mountain about an eighth of a mile distant forms a clearly cut profile, known as the **Iron Duke.**   Taking a path in that direction, thirty rods will reach a point of the ledge called **Dramatic Rock.**   A few feet beyond is a fissure in the ledge some thirty-five feet deep and eight feet wide—**Styles' Gorge**—crossed by a rustic foot-bridge, named **Grace's Bridge.**   The **Devil's Kitchen,** a triangular hole in the rock thirty feet deep, is soon passed, and **Wellington's Rock** reached.   This rock is a large bowlder with its inner edge resting on the ledge and its outer supported by a column of rock split from the side of the rocky wall.   It is the Iron Duke's cap.   A ladder reaches to the top, from where a wonderful view is obtained of the immense depth immediately below and of the Hudson and Esopus valleys.

Just beyond is **Pulpit Rock.**   Here the projecting face of the mountain descends perpendicularly about a hundred feet.   Next is **Overlook Ledge** where a path descends a ridge of the mountain, called **Minister's Face,** winding down a succession of ledges to the **Dominie's Nose,** where the mountain is almost perpendicular for a depth of five hundred feet. The spot is a mile from the house.

The path from Overlook Ledge on around the summit of the mountain soon comes to **Bishop's Rock.** Near, to the north are Plaaterkill Mountain and Indian Head; beyond them stand High Peak and Round Top; and still farther away are Black Head and Black Dome. Kingston, Catskill, Hudson, and Albany are visible. From here the path turns to the left, and a few yards along comes to **Turtle Rock**—a bowlder whose top is a fair representation of a turtle. (From this point a path turns down the mountain to the road to Plaaterkill, reaching it a fourth of a mile distant.)

As the path rounds the north side of the mountain to return there is a point of the ledge, called **Kimball's Rock,** which discloses a beautiful view of Echo Lake, about three-fourths of a mile to the north and a thousand feet below. **Glen Evans** is on the return path—a cool recess among huge blocks of rock, detached from the western side of the mountain top. Not far from here is **Hawkin's Rock,** which affords the best view of the succession of mountain peaks to the north and west.

There are other interesting places on the summit of the mountain, all within half a mile of the Overlook House.

MINNEHAHA SPRING.

WALK IN THE OVERLOOK MT. HOUSE REGION.

Minnehaha Spring is a fourth of a mile west of the house. Remarkably clear water flows from underneath a ledge and is caught in a pocket of the rock, making it convenient to drink therefrom, before it is lost in the forest below.

ECHO LAKE.

WALK OR DRIVE IN THE OVERLOOK MT. HOUSE REGION.

Echo Lake, or, **Shuc's Lake,** as it was formerly called, is two miles north from the Overlook House, near the new road between this house and Plaaterkill. A branch road leads to it. The Lake is very deep and the water clear and cold. Mountain tops covered with forest rise around, a thousand feet above its level. It is fed by rills and springs and discharges its waters into the Sawkill to the south-west. Trout are found here.

NOTE.—Mention is made in this Guide of a new road between the Overlook Mountain House and Plaaterkill. That was written and put in type while the construction of the road was being carried on. The work was finally suspended when the road was finished half way, and so it is only complete from the Overlook House to Echo Lake. It will probably be finished for next season. There is a path the whole distance.

CAIRO.

The village of **Cairo** is on high land near the north-eastern base of the Catskills. The route to it is by stage from Cats-kill, ten miles distant. The view of the mountains from this near point is excellent. Telegraphic communication.

HENSONVILLE.

Hensonville village is a pleasant spot among the mount-ains at an elevation of some fifteen hundred feet. It may be reached by stage from Catskill, twenty-five miles, or, by stage from Phœnicia, twenty miles and a half. To the east, at the head of the valley in which this place is located, rises the Black Dome mountain to an altitude of 4,003 feet. It is two miles and a half from Windham. Telegraphic facilities.

WINDHAM.

Windham is as beautiful a village as the county affords. It lies in a mountain valley at an elevation of 1,510. A long ridge of mountain peaks rises on either side. To the east is Windham High Peak which has an altitude of 3,534 feet, and affords an extended view from its summit.

The place was settled in 1785. There are churches, two telegraph offices, and a weekly newspaper. The few boarding houses are very pleasantly located. The way to reach this place is by stages either via Catskill, twenty-six miles; Phœ-nicia, twenty-three miles; or, Prattsville, ten miles. The first route, although a longer stage ride, is the most direct.

PROSPECT PARK HOTEL,
CATSKILL, N. Y

(See illustration in description of Catskill.)

FIRST-CLASS SUMMER RESORT,

Of Easy Access on the Banks of the Hudson River,

WITH ALL THE LATEST IMPROVEMENTS.

The main building is two hundred and fifty feet front by forty feet, with wing one hundred and forty by forty feet. Dining-room full length of wing ; with two-story Piazza three hundred and seventy by sixteen feet.

The grounds, walks, avenues and shrubbery are adapted to the chief design : which is, to produce such an establishment, on a liberal and appropriate scale, as can offer to those who with their families annually seek in the country, during the Summer months, health and grateful change from the heat and confinement of the city. No malaria, hay fever or mosquitoes. Croquet, Billiards, Bowling Alley, Fishing, Boating, Bathing, good Music.

The Views from the Hotel are Unsurpassed in Extent and Beauty.

The annually increasing tide of visitors to this region—drawn hither in the pursuit of health and pleasure – has already vindicated its right to the title of "The Switzerland of America."

The location is a judicious selection from the Prospect Hill ; and the site, with its surroundings, occupies twenty acres. The plateau is admirably adapted to the purpose. With a commanding view of the River in front, and for miles north and south. and the grand old Mountains in the background, with a climate of great salubrity, healthy mountain air, and the accessories of field and river sports and pleasure drives, it is unsurpassed in all the borders of the Hudson in its attractions and advantages.

Carriages will be in attendance at the Cars and Boats.

Accessible by nearly all trains on the Hudson River Railroad, and by the Day Boats "Chauncey Vibbard" and "Daniel Drew." Also by Night Steamers every evening from foot of Harrison Street, New York, for Catskill.

First-class Livery connected with the Hotel, with good stabling for horses in new brick stables recently erected.

Western Union Telegraph in the House.

PRICES TO SUIT THE TIMES.

Address PROSPECT PARK HOTEL CO.,

Catskill, **N. Y.**

IRVING HOUSE
CATSKILL, N. Y.

Is a large, new and commodious brick building. in the centre of the Village, with

First-Class Accommodations

FOR TRAVELERS AND TOURISTS.

The Village Omnibus and Baggage Wagon attend all Trains and Boats.

An authorized Agent will be on hand to give information, accommodate and attend to the wants of the guests of this House.

Parties desiring Country Board, furnished with information regarding the different localities and prices.

H. A. PERSON, Proprietor.

SUMMIT HILL HOUSE
CATSKILL, N. Y.

Twenty minutes' walk from Depot and Steamboat Landing. Large, well ventilated rooms. Capacity, a. On a large farm using our own Fruit, Vegetables Milk, &c.

P. M. GOETCHIUS,   -   PROPRIETOR.

MAPLE GROVE HOUSE,

Entrance to Cauterskill Clove.

————•—•—————

Located amidst the Beauties of the Catskills.

House Enlarged and Refurnished.

Billiards, Bowling, and good Stabling.

For terms, address   PHILO PECK,

Palenville, Greene Co.. N. Y.

PALENVILLE HOTEL,

PETER BURGER, Proprietor.

Near the First Bridge in the Cauterskill Clove.

————•—•—————

A very Pleasant Location and Favorite Resort.

Comfortable Rooms.   Excellent Table.   Terms, $7 to $10.

Watering Tank for Horses.

HENSONVILLE.

GRIFFIN'S RURAL RETREAT.

Situated in a Pleasant Hamlet of the Catskills ; Two Telegraph offices and Post-office within a stone's throw of the House.   Fine accommodation for drives to places of interest. House New, and newly furnished.   Terms low.

For particulars address   O. S. GRIFFIN, Prop'r,

Hensonville, N. Y.

THE MOUNTAIN HOME,

Tannersville, Greene County, N. Y.

AARON ROGGEN, Proprietor.   WILL P. ELLIS, Clerk.

This house is located on the Catskill Mountains, 15 miles from Catskill, accessible by a daily line of Mail stages.

Good Trout Fishing in the Season.

Conveyances on the premises for Pleasure Parties.

Post-office and Telegraph Office in the House.

For further particulars apply to MORE, NICOLL & FITCH, No. 7 Warren St., or THOMAS GROVES, with A. T. Stewart & Co., New York.

Tannersville Cottage,

Tannersville, Greene Co., N. Y

NEW HOUSE WITH LARGE AND AIRY ROOMS, ON THE CATS-KILL MOUTAINS, 14 MILES FROM CATSKILL LANDING.

Accessible by a Daily Line of Mail Stages.

Post-office and Telegraph Office are but two minutes' walk distant.

Former patrons of this house have been well suited, and no pains will be spared to continue to make this a pleasant Summer Resort.   Accommodations for 20, at reasonable terms.

GEO. CAMPBELL, PROPRIETOR.

ELM COTTAGE,

Tannersville, Greene County, N. Y.

THIS HOUSE is situated in a pleasant spot on the Mountains, 15 miles from Catskill, and has accommodations for fifteen people.   It will be the endeavor to continue to give satisfaction to its patrons.   The Post-office and Telegraph Office are within five minutes' walk.   Terms and full particulars furnished on application.   MISS L. A. CRAIG.

CASCADE HOUSE,

Tannersville, Greene County, N. Y.

This House is located on the western slope of the Catskill Mountains; 15 miles from Catskill, and 6 miles from Catskill Mountain House. Accessible by daily stages. Post-office and Telegraph Office within 15 rods of the House.

TERMS VERY MODERATE.

Conveyances on the premises for parties of pleasure, and meeting parties at the Boats and Cars.

G. N. EGGLESTON, Proprietor.

MEADOW-BROOK HOUSE,

Catskill Mountains, 15 Miles from Catskill.

SUMMER BOARD AT REASONABLE RATES.

Carriages and Horses on the Premises.

This House is pleasantly located on the western slope of the Mountains, near a delightful meadow brook, with cheerful surroundings.

Address A. STIMPSON HAINES,

Tannersville, Greene Co., N. Y.

FOUR-HORSE STAGES

—FOR—

PALENVILLE, TANNERSVILLE
HUNTER AND LEXINGTON

Leave Catskill Daily,

Connecting with Catskill Boats Each Way.

GILBERT HAINES, Proprietor.

MAPLEWOOD.

This long established House is located amid the chief attractions of the Catskills at an elevation of about 2,000 feet. Some large, airy rooms have recently been added, and it has comfortable accommodations for 25 people.

Daily Mail. Post-office and Telegraph Office near.

Terms and particulars, and best of references in New York, Philadelphia, Boston and other cities, furnished.

E. H. LAYMAN, Tannersville, N. Y.

GLEN COTTAGE,
CATSKILL MOUNTAINS.

This Summer Resort is capable of accommodating 45 guests. Located in a quiet, romantic neighborhood, amidst beautiful mountain scenery. Elevation 2,500 feet. Directly opposite High Peak and Round Top; four miles west of the Mountain House, three miles from the Laurel House or Cauterskill Falls, one and a half miles from Haines' Falls, and fifteen miles from the landing at Catskill. Accessible by H. R. R., Albany Day Boats, and Catskill Steamboats, to Catskill, where a conveyance will be in waiting.

OWEN GLENNON, Proprietor.

Post office Address, Catskill, Greene County, N. Y.

Telegraph Address, Haines' Falls, N. Y.

ROBERT KERR'S
DAILY STAGE

For Palenville and the Haines' Falls Region.

Meets the Boats and Trains at Catskill. Always in waiting upon arrival of the Day Boats.

Only authorized convevance for THE VISTA, the next house to the Haines' Falls House.

Rusk's Illustrated Guide to the Catskill Mountains; with Maps and Plans, 75 cents. Guide, complete, without Guyot's Map, 25 cents.

Guyot's Map of the Catskills, covers, pocket form, 75 cents.

Rusk's Map of the Heart of the Catskill Mountains, covers, pocket form, 25 cents.

The above works are for sale at numerous places in the Catskills. Mailed upon receipt of price.

SAMUEL E. RUSK, Catskill, N. Y.

HUNTER HOUSE,

Hunter, Greene County, N.Y.

This Hotel is now Open for the Accommodation of Boarders and the Public Generally.

Daily Line of Stages between Hunter and Phœnicia, connecting with all trains to and from New York.

Parties met with private conveyance if desired.

M. C. VAN PELT.

FURNITURE.

D. B. BALDWIN & CO.,

HUNTER, N. Y.,

Dealers in

Parlor, Chamber and Dining Room

FURNITURE,

BEDDING, SPRING BEDS, &c.

We make a specialty of Furnishing Hotels and Boarding Houses.

CENTRAL HOUSE,

HUNTER,

Greene County, New York.

(See illustration in description of Hunter.)

THIS-house is located in the beautiful Village of Hunter, in the heart of the Catskills. It has ample grounds and plenty of shade. Piazzas about 150 feet in length. Pure spring water on each floor. No bar. The locality affords freedom from chills and fever, malaria and hay fever.

Hunter Mountain—the highest of the Catskills, 4040 feet and the rugged spur, the Colonel's Chair, rise directly in front. (See map, in this book.) Among the drives, over good roads, may be mentioned, to the Catskill Mountain House, Kaaterskill Falls, Haines' Falls, Plaaterkill, Tannersville, Stony Clove, Grand View and Windham. There are many well shaded walks to places of interest.

The most direct route to Hunter is via Hudson River to Rondout ; thence by rail to Phœnicia, and by stage through the renowned Stony Clove, where ice remains even in mid Summer. Private conveyances will meet parties at Phœnicia if desired. Daily mail between here and New York. Post-office, next door. Two Telegraph Offices. Two Physicians. Three Churches.

Circulars giving full information and reference to former patrons furnished on application. Address the proprietors,

J. RUSK & SON.

THE EXAMINER,

ISSUED EVERY SATURDAY MORNING. FIFTY-SECOND YEAR OF PUBLICATION.

M. H. TROWBRIDGE, Editor and Proprietor.

A large nine-column paper containing all the local county and summer resort news. Arrivals at the leading hotels also reported.

TERMS, - **$1.50 in advance.**

The Steam Job Printing Department

Of the Office is complete in all respects.

Office in Martin's Building. 301 Main St.

CATSKILL, N. Y.

THE COXSACKIE NEWS,

A WEEKLY NEWSPAPER, *ESTABLISHED IN 1867*

Contains a full report of all news, both foreign and local. Especial attention given to the news of Greene county.

NOT THE BEST PAPER

In the county, perhaps, but taking *no back seat* for any other.

A GOOD ADVERTISING MEDIUM, AND RATES MODERATE.

A First-class Job Printing Department is run with the Paper, and new type and power presses put us on a good business footing with our cotemporaries.

Terms of Subscription to NEWS, $1.50 *per Annum, strictly in advance.* 75 *cents for six months.*

WM. P. FRANKLIN.   S. M. AUSTIN.

THE
AMERICAN GUIDE BOOKS.

New England: with 6 Maps and 11 City Plans.

Before you begin to travel in New England, be sure to provide yourself with Sweetser's "Hand Book." It is a small compact volume, with maps and plans and tours ; with history condensed, and such illustrations as make it a constant help and pleasure to the tourist. It is admirably put together and is a vast labor-saving guide for one who wishes to know what to see and what he is seeing.—Rev. Dr. PRIME, in *New York Observer.*

It is about as nearly faultless as such a book can be.—*New York Tribune.*

The book is compact and crowded * * * The information in regard to the different localities is full, minute and exact.—*Boston Transcript.*

It is by all odds the best book of the kind that ever has appeared, leaving all others far behind—so far, indeed, that they are out of sight of it.—*Boston Traveler.*

The Middle States : with 8 Maps and 15 City Plans.

No previous manual is so copious or so exact in its treatment, or can be consulted to so great advantage by the tourist in the Middle States as a trustworthy guide.—*New York Tribune.*

The maps alone are worth the price of the volume, which is crammed with knowledge like a traveler's valise with luggage.—*Daily Graphic.*

The work is very faithfully done, and the 500 pages are crammed with facts useful to the tourist.—*Springfield Republican.*

The Maritime Provinces : with 4 Maps and 4 City Plans.

By its intrinsic value, copiousness of information, and impartiality, it is likely to take the place of all other guides or handbooks of Canada which we know of.—*Quebec Chronicle.*

In graphic and picturesque description, in completeness and fullness of information, and in clear insight into a traveler's needs and perplexities, this guide book is not to be excelled.—*Boston Journal.*

The White Mountains : with 6 Plans and 6 Panoramas.

Altogether, in plan and workmanship, this guide-book is as perfect a thing of its kind as could well be produced. It is simply indispensable to all who visit or sojourn among the White Mountains.—*The Congregationalist.*

By far the best guide through that favorite region of summer tourists that has yet been published. Indeed, the book combines all the information that any intelligent being can possibly need for making a thorough exploration of the White Mountain country, on foot, by rail, by stage or carriage.—*Phila. Bulletin.*

** Price $2,00 each.   For Sale by Booksellers.

Sent by mail on receipt of price by the Publishers,

HOUGHTON, OSGOOD & CO., Boston.

BLYTHEWOOD.

THIS SUMMER RESORT, beautifully situated in the heart of the Catskill Mountains, will be open for Boarders the 1st of June. It is accessible from Catskill village, by two daily lines of stages. The house is new, has all modern conveniences, and is comfortably furnished throughout There is abundant shade, fine lawn and croquet ground on the premises. Good trout fishing in the neighborhood during the season.

Comfortable private conveyances can always be furnished for pleasure parties, and will be sent to meet guests on the arrival of cars or steamboat, when desired.

Post-office and Telegraph Station within half a mile of the house.

MRS. ALEX. HELMSLEY,
Tannersville, Greene Co., N. Y.

FAIRMOUNT HOUSE,
TANNERSVILLE, GREENE CO., N. Y.
WILLIAM WOODEN, Proprietor.

This is a new house, with large and airy rooms ; located in the western slope of the Catskills, sixteen miles from Catskill Landing, and six miles west of the Catskill Mountain House. Daily stages from Catskill. Telegraph and Post-office one-fourth mile distant. Conveyances for pleasure parties.

For further particulars inquire of Rev. A. Crosby, 96 Orange St., Brooklyn.

JACOB FROMER,
TANNERSVILLE,
SELLS

Zephyr Worsteds,
Germantown Yarns,
Hosiery, Boots and Shoes,
AND A
GENERAL ASSORTMENT OF NOTIONS AND DRY GOODS.

CONFECTIONERY, FRESH FRUITS, DRUGS AND GROCERIES.

Store but a few steps east of the Post-office.

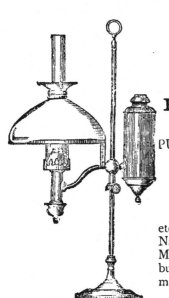

BEST IN THE WORLD!

Common-Sense Chairs and Rockers,

With or without Reading Table.

No Office, Library (public or private), Sitting-room, or Piazza. should be without some of my Rocking Chairs ; so roomy, so easy and durable. Try my Puritan Rocker, or Old Point Comfort, and you will find *Rest*.

My Reading and Writing Table is separate from chair, and is secured in position by a strong button. Is easily adjusted to nearly all kinds of arm-chairs, but should be used on my No. 4, 5, or 13, to give *complete* satisfaction. Table can be set at any angle desired, or lowered to good position for writing. Makes a nice table for an invalid. Cutting-board for the ladies. None of your little 7x9 affairs, but is 16x32 inches. Cannot be got out of order.

FOR SALE BY THE TRADE.

MANUFACTURED BY

F. A. SINCLAIR, Mottville N. Y.

Send stamp for Illustrated Price List.

Shipped as freight or expressed to all parts of the United States.

ROGERS' GROUPS
$10 and Upward.

THE PEDDLER AT THE FAIR.

The Peddler is on horseback, with his box of jewelry before him, and watches with interest the result of the solicitations of the young lady by his side, who is coaxing her father to buy a necklace.

These Groups are Packed to go with safety to any part of the World.

[See opposite page.

ROGERS' GROUPS.

HIDE AND SEEK, (Boy)	$50 00
HIDE AND SEEK, (Girl),	50 00
PEDESTAL—for Hide and Seek—each	10 00
BUBBLES	35 00
FAIRY'S WHISPER	25 00
FUGITIVE'S STORY	20 00
COUNCIL OF WAR	20 00
THE MOCK TRIAL	20 00
CHALLENGING THE UNION VOTE	20 00
POLO	15 00
THE PHOTOGRAPH—THE PAIR	15 00
THE PEDDLER AT THE FAIR	15 00
TAKING THE OATH	15 00
THE FAVORED SCHOLAR	15 00
PRIVATE THEATRICALS	15 00
THE TRAVELING MAGICIAN	15 00
WEIGHING THE BABY	15 00
CHECKERS UP AT THE FARM	15 00
TAP ON THE WINDOW	15 00
WASHINGTON	15 00
THE FOUNDLING	15 00
COMING TO THE PARSON	15 00
COURTSHIP IN SLEEPY HOLLOW	15 00
ONE MORE SHOT	15 00
WOUNDED SCOUT	15 00
UNION REFUGEES	15 00
COUNTRY POST-OFFICE	15 00
SCHOOL EXAMINATION	15 00
CHARITY PATIENT	15 00
UNCLE NED'S SCHOOL	15 00
RETURNED VOLUNTEER	15 00
PLAYING DOCTOR	15 00
SCHOOL DAYS	12 00
PARTING PROMISE	12 00
RIP VAN WINKLE AT HOME	12 00
RIP VAN WINKLE ON THE MOUNTAIN	12 00
RIP VAN WINKLE RETURNED	12 00
WE BOYS	12 00
MAIL DAY	10 00
TOWN PUMP	10 00
PICKET GUARD	10 00
GOING FOR THE COWS	10 00
THE SHAUGHRAUN AND "TATTERS"	10 00
HOME GUARD	10 00

ILLUSTRATED CATALOGUES may be had on application, or will be mailed by enclosing ten cents to

JOHN ROGERS,

23 Union Square, (formerly 1155 Broadway,) New York.

Please state where you saw this Advertisement.

THE SCIENTIFIC AMERICAN.

. *THIRTY-FOURTH YEAR.*

THE MOST POPULAR SCIENTIFIC PAPER IN THE WORLD.

Only $3.20 a Year, including Postage. Weekly. 52 Numbers a year 4,000 book pages. The SCIENTIFIC AMERICAN is a large First Class Weekly Newspaper of Sixteen Pages, printed in the most beautiful style, *profusely illustrated with splendid engravings,* representing the newest Inventions and the most recent Advances in the Arts and Sciences ; including New and Interesting Facts in Agriculture, Horticulture, the Home, Health, Medical Progress, Social Science, Natural History, Geology, Astronomy. The most valuable practical papers, by eminent writers in all departments of Science, will be found in the Scientific American ;

Terms, $3.20 per year, $1.60 half year, which includes postage. Discount to Agents. Single copies ten cents. Sold by all Newsdealers. Remit by postal order to MUNN & CO., Publishers, 37 Park Row, New York.

PATENTS. In connection with the Scientific American, Messrs. MUNN & CO. are Solicitors of American and Foreign Patents, have had 34 years experience, and now have the largest establishment in the world. Patents are obtained on the best terms. A special notice is made in the Scientific American of all Inventions patented through this Agency, with the name and residence of the Patentee. By the immense circulation thus given, public attention is directed to the merits of the new patent, and sales or introduction often easily effected.

Any person who has made a new discovery or invention, can ascertain, free of charge, whether a patent can probably be obtained, by writing to the under signed. We also send *free* our Hand Book about the Patent Laws, Patents, Caveats, Trade Marks, their costs, and how procured, with hints for procuring advances on inventions. Address for the Paper, or concerning Patents,

MUNN & CO., 37 Park Row, New York.

Branch Office, Cor F & 7th Sts., Washington, D. C.

ESTABLISHED 1850.

Studio 241 Warren Street, Hudson, N. Y.

PHOTOGRAPHY, in all its branches, Carbon Photographs, Transparencies and Porcelains—guaranteed to be unalterable. This is the only Gallery between New York and Albany where these pictures are made. Frames, Albums, Passepartouts, Engravings, Chromos, &c. Latest styles and Publications always on hand. ·

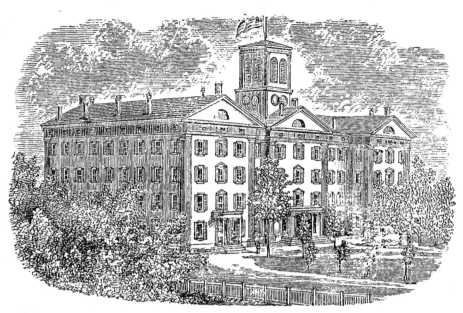

Claverack College and Hudson River Institute,

REV. ALONZO FLACK, PH. D., PRESIDENT,

Is one of the largest and best sustained Boarding Institutions for both sexes in this country, and is located in the village of

Claverack, Columbia County, New York,

Three miles from Hudson and eight from Catskill, commanding a fine view of the Catskill Mountains and Mountain House.

The Building contains 240 Rooms : comprising 146 furnished rooms, accommodating two pupils each; 13 Professors' and Teachers' rooms; 12 Lecture and Recitation, 28 Piano, and 4 Literary Society and Reading rooms ; a Library, an Armory; a Drill Hall and Gymnasium; a Chapel; 2 Offices; and 35 rooms for Domestics, and Domestic and Culinary purposes.

The Institution has a fine set of Chemical and Philosophical Apparatus, and a large Library.

There are Eleven Departments of Instruction : English, Normal, Classical, French, German, Musical, Painting, Military, Commercial, Agricultural and Telegraphic. Each department receives the especial attention of the instructor in charge.

In addition to the general Academic and special courses of study, there is a Collegiate Course for Women, prescribed by the Board of Regents, which entitles those having completed it to the degree of Batchelor of Arts.

For Catalogues, apply to

ALONZO FLACK, President.

ITHACA
CALENDAR CLOCK.

A Perpetual Mechanical Calendar connected with the most Superior Eight and Thirty-day (either weight or spring) Clock Movements.

IT INDICATES PERPETUALLY

The Day of the Month,
The Month of the Year.
The Hour of the Day,
The Day of the Week.

Calendars Printed in the English, Spanish, Portugese, French, German, Russian, Turkish, and Asiatic Languages.

It is indispensable to every place of business—a necessity in every household. Manufactured in numerous styles, ranging widely in prices to suit the various wants of the public. All clocks are thoroughly regulated, and calendars mechanically worked through the changes of eight (8) years of time before leaving the manufactory. For sale by all leading jewelers in the United States and Canada.

Catalogues and Price Lists mailed, and inquiries answered promptly on application.

Ithaca Calendar Clock Company,

Manufactory at ITHACA, N. Y.

New York Office with Waterbury Clock Co., No. 4 Cortlandt Street.

INDEX.

THE

LAUREL HOUSE,

AT THE KAATERSKILL FALLS,

Catskill Mountains, 1½ Miles West of Mountain House.

J. L. SCHUTT, Proprietor.

This new and spacious Hotel, recently enlarged and re-furnished, is located at the celebrated KAATERSKILL FALLS, on the eastern summit of the Catskills.

The Falls have been well described by BRYANT and COLE, and by COOPER in The Pioneers. The first Fall is nearly two hundred feet high, and the water looks like flakes of snow as it strikes in the pool below. Working along the rocky shelf, it falls another hundred feet, and then descends the wooded glen in a succession of cascades.

In the immense rocky amphitheater which sweeps like mason work in the rear of the first Fall, are paths on which the visitor may pass entirely around behind the falling water. Through the wide ravine may be seen the western side of High Peak and the adjacent mountains. The walks in the vicinity include those to Sunset Rock, North and South Mountains, and Haines' Falls, and there are pleasant drives through the Cauterskill Clove and over the adjacent mountains.

Good hunting and trout fishing in the neighborhood.

A wing, 50x50 feet, has recently been added to the House, greatly extending the accommodations at this popular resort. Carriages, Stages and an authorized Agent in attendance at the Cars and Boats, Catskill.

red and re-fur-
ILL FALLS, on

VT and COLE,
is nearly two
es of snow as it
rocky shelf, it
he wooded glen

weeps like ma-
on which the
falling water.
rn side of High
in the vicinity
uth Mountains,
es through the
ains.

borhood.

to the House,
popular resort.
attendance at

LAUREL HOUSE KAATERSKILL FALLS

FRANCE'S GREATEST ORGANIST.

In France to-day the highest ambition of the organist, as well as the crowning hon-or which can be conferred upon him, is the appointment of "Organist of the Church of the Madeleine, Paris." It is with no small satisfaction, therefore, that Messrs. ESTEY & Co. present the following testimonial from the present occupant of that proud position, far outranking any decision of juries :

"I have played upon the organs of Messrs. ESTEY & Co., and have been charmed with their quality of tone, which comes very near that of a pipe organ, and also with the resources it gives to the player.   CAMILLE DE SAINT SAENS.

CPSIA information can be obtained
at www.ICGtesting.com
Printed in the USA
LVOW13s2104130318
569767LV00023B/290/P